Dedication

To our parents
Una, Ted, Elizabeth and Lyn
whose support has been inspirational.

make the dream real

Donald Douglas Snr, the aviation industry's and one of the world's greatest industrialists poses with the DC-4. This was his favorite photo.

Dream no small dream; it lacks magic.
Dream large. Then make the dream real.
– Donald Douglas Snr —

FLIGHTPATHS - Exposing the myths about airlines and airfares
By Geoffrey Thomas and Christine Forbes Smith

2nd edition published in May 2007 by
Aerospace Technical Publications International Pty Ltd
P.O. Box 8205,
Perth Business Centre,
Stirling Street Post Office, 6849,
Western Australia
geoffreythomas@iinet.net.au
www.flightpaths.com.au
Tel: + 61 41 793 6610
Fax: +61 8 9309 5991

Authors: Geoffrey Thomas and Christine Forbes Smith
Editor: Christine Forbes Smith
Art Work/Graphs: Juanita Franzi - Aero Art Publications

Photographers: Art Brett, Mark Garfinkel, Craig Murray, Geoffrey Thomas, Captain Kevin Tate, Emmanuel Tailliet, Misha Popov (www.mishapopov.com)
Style Editor: Kerry Coyle

Design Concept: Kym Weir-Smith and Jim Davies - 303 Advertising, Perth.
Additional layouts and Concepts: Todd Shaw, Perth W.A
Printed by: APOL China

ISBN: 978-0-9752341-4-3

Contents

Acknowledgments

It is impossible to put into words the gratitude we feel to those who have helped bring "Flighpaths" to lift-off. The assistance has ranged from words of support from industry leaders such as Rod Eddington (Former Chief Executive of British Airways) to tireless support from Ken Morton, (Communications Director Australia/NZ, The Boeing Company) and his PA, Emma Hodsdon. Always ready to assist have been Ted Porter (Aviation - Aerospace Media Consultants - Australia's Airbus media relations), and Françoise Maenhaut (Media Relations Coordinator Communications at Airbus).

Great encouragement, friendship and material have come from dear friends of many years Jon Proctor and Mike Machat. Always ready, willing and able to help over the years of research have been: Harold Adams (Chief Designer (ret) Douglas Aircraft Company); the late Don Hanson (former VP of Douglas Aircraft Company Corporate Communications); the late Dewey Smith who took many of the striking air-to-air images of Douglas aircraft; Patricia McGinnis (Boeing Historical Archives - Long Beach); Yvonne Leach (787 Communications); David Doorman, (BAe Systems); Paul Jarvis and Bob Petrie (British Airways Museum of Flight) and Alan Renga (San Diego Aerospace Museum).

The following organisations have also provided data or photographs over many years that have been used in this book: AirAsia, Air Transport Association, American Airlines, British Airways, Cathay Pacific, Emirates, Honeywell, IATA, ICAO, Lockheed, Lufthansa, Pan Am, Singapore Airlines, Western Airlines and TWA.

Finally, we would like to thank our five wonderful boys, Christopher, Nicholas, Alex, Simon and Mitchell for their patience and understanding while this project has taken so much of our time and focus over the years.

Geoffrey Thomas &
Christine Forbes Smith

About the authors

Geoffrey Thomas is the Senior Editor of the world's leading airline management journal Air Transport World and has been studying, writing about and commenting on commercial aviation for 35 years. Mr Thomas is based in Perth, Australia and was previously SE-Asian Contributing Editor for Aviation Week and Space Technology.

He was named Aerospace Journalist of the Year in the Best Systems or Technology Submission for 2002 and 2003 and was named Australasian Journalist of the Year for 2001 and 2002. In all Mr Thomas has won 16 international and Australasian awards.

He has been published in The West Australian, Sydney Morning Herald, The Age, The New Zealand Herald, South China Morning Post, Sunday Times and Sunday Telegraph.

Mr Thomas is a regular commentator on Australian and New Zealand radio and television. Flightpaths is Mr Thomas' sixth book.

Christine Forbes Smith (B.A. (Social Sciences) and B. App. Sci.) is regularly published in W.A Business News, and has been published in The West Australian travel pages focusing on consumer issues as well as destinations. This is Christine's sixth book.

Introduction

Possibly nothing evokes as much discussion and yet so much misunderstanding as the cost of air travel today. The perception is that airfares are sky-high, yet the reality, as Flightpaths, will show, is that they have never been so cheap regardless of how far you fly.

Whichever comparison you use – the rise in inflation, average weekly earnings, CPI, the cost of housing, fuel, bread, milk or the sticker price on the family car – air travel has demonstrated the least significant price increase. Air travel's "high cost" is just one of the many urban myths about the airline industry put to rest in Flightpaths – the first book to explain in simple terms the complex link between technology and airfares.

Technology's effect on aviation has been pervasive and the industry is littered with the wreckage of airlines and aircraft manufacturers that miscalculated or just didn't fully understand what a pivotal role technology would play in driving down costs and improving the performance of aircraft. Similarly, airline unions have largely failed to grasp the worldwide trends in aviation.

The phenomenal growth and lack of profitability in the airline industry is consistently underestimated by authorities and governments, placing immense strain on the industry's infrastructure, which for the air traveler manifests itself in long airport check-in queues and for airlines in queues to take-off or land.

While it is not necessary to understand why a Rolls-Royce Trent jet engine can run at temperatures as high 1,600C - 300C higher than the melting point of its turbine blades – it is vital for them to understand the result - lower fares and shorter flight times.

To illustrate the growth of the industry and the dramatic decline in airfares, the book focuses particularly on the US and Australian markets because they have been largely unaffected by the ravages of war and both are mature markets.

The US and Australia have also been at the forefront of deregulation, liberalization and open skies policies, while Australia's Qantas is the world's second oldest airline and a leader in technical innovation.

There is also comprehensive data available back to the 1930s, essential to make detailed historical comparisons to illustrate the stunning advances in aviation. For similar reasons of continuity, Rolls-Royce, which has been at the cutting edge of engine technology, is the focus of much of the comparative engine data. From a manufacturing aspect Douglas, Boeing and Airbus have dominated the airline industry and thus the advancements made by these companies are highlighted.

Then-year dollar comparisons have been used throughout and where a current perspective is required we have converted the figure to 2003 dollars. The US dollar has been used as it is the international standard in aviation, however graphs 15.1 and 15.6 are shown in AU dollars. Imperial and metric measurements are shown.

Flightpaths will take you on a fascinating journey of discovery into the world's most misunderstood yet vital industry. The book is, we believe, a magnificent photographic work that not only visually reinforces key points but is also a celebration of flight through the camera.

We sincerely hope that you enjoy Flightpaths and we welcome your feedback.

Geoffrey Thomas & Christine Forbes Smith

CHAPTER 1

Early Years

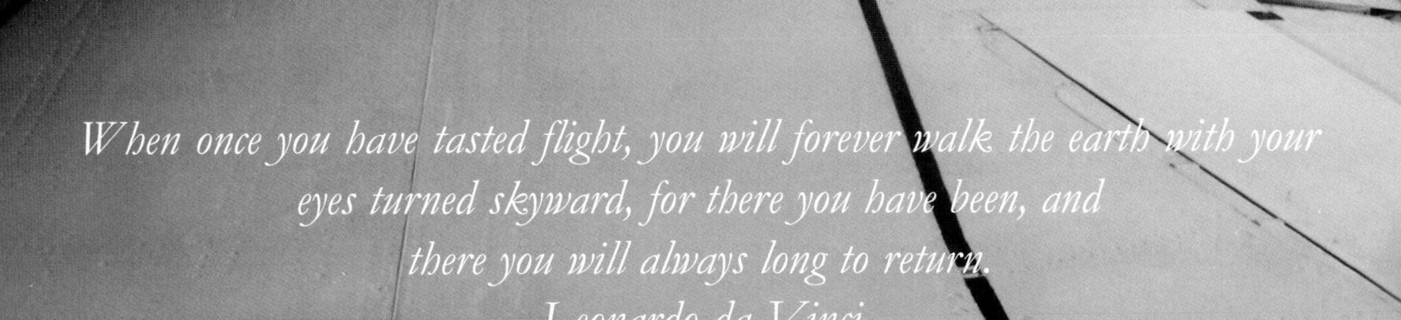

When once you have tasted flight, you will forever walk the earth with your eyes turned skyward, for there you have been, and there you will always long to return.

— Leonardo da Vinci —

EARLY PIONEERS AND FLYING MACHINES

The first aircraft to take to the air were judged by many so-called experts principally for their novelty or entertainment value. Rash and often comical attempts were made to become airborne with a beneficial result that many principles of aerodynamics were refined. As early as 1911, light but useful loads were being lifted in such countries as India, England, Italy and the US.

HP W.10 was a development of the WW1 bomber and could carry 16 passengers.
British Airways Museum

The progress in Russia, however, was nothing short of spectacular, with Igor Sikorsky building his "Grand" and "Ilya Mourometz" which in 1913, carried 12 passengers on a six-hour flight. Incredibly, in a first and last for aviation, the passengers on this flight were able to stroll outside on open balconies and the first in-flight meals were served.

In 1914, there was one attempt in the US to start a regular passenger service across Tampa Bay. Only one passenger could be carried and the cost for the journey was $5 for the 20 mile (37.04 km) journey - one-third of the average weekly earnings at the time. Generally, early development was slow.
With the advent of World War I however, this changed dramatically as the potential for aircraft to carry large bomb loads was recognized.

A West Australian Airways DH-66 Hercules "City Of Perth" preparing for take-off in Perth, Western Australia in 1929. Note female passenger leaning out of the window, just near the engine. *Ansett Australia*

DON'T THROW FOOD OUT OF THE WINDOW

As a result of the spectacular success of aircraft during the war effort, a host of countries set up air services to restore and build on pre-war mail services. Only a few saw the merit in carrying passengers. One such airline was Lignes Aeriennes Farman, based in France, which used a Farman Goliath to carry passengers from Paris to Brussels in March 1919.

Across the channel in England, aircraft manufacturer A.V. Roe and Company launched an air service from Manchester to Blackpool using a three-seater Avro aircraft. But the development of the British airline industry was severely hampered by the decision of the then Secretary of State Winston Churchill, who declared that airlines should "fly by themselves", thereby eliminating the government as a possible source of financial help.

In stark contrast, European governments were aiding their airlines with various subsidies - a situation that has only really changed in the past 10 years. Despite the lack of government support, Aircraft Transport and Travel, founded by one of the era's great supporters of air travel, George Holt Thomas, started a daily international service between the UK and France. The proving flight on 15 July 1919 carried one passenger, the influential Major Pilkington of the famous glass manufacturing firm. The fare for the two-hour and 30-minute journey was 21 English Pounds (£), a king's ransom in those days.

Flying was only for the rich and/or brave and the regularity of flights was not assured. Adding to the woes of flying at that time, passenger cabins were not heated, pilots were seated in open cockpits and navigation was entirely visual. Ventilation wasn't a problem because windows could be opened, which prompted one airline to put up a sign asking passengers not to throw food out the window. Most airlines followed railway lines, some railway stations assisting by painting the name of the stations on their roofs. Additionally, because engines were so unreliable a host of emergency landing fields were set up on the various routes.

Boeing Monomial 220 was designed to carry mail but a larger version also carried six passengers. *Boeing Historical Archives*

THE BRAVE'S QUEST FOR DISTANCE, FAME AND FORTUNE

The story of the progress of airlines and aircraft is also the story of those who challenged the unknown. Helping to give them courage to risk their lives over great stretches of ocean were newspaper magnates and governments. One of these was Britain's Lord Northcliffe, who four days after the close of WWI offered £10,000 for the first non-stop trans-Atlantic flight.

John Alcock and Arthur Whitten-Brown were the first to cross the Atlantic on 15 June 1919 in just over 16 hours in a Vickers Vimy - a WWI twin-engine bomber. That aircraft was also the choice for brothers, Captain Ross and Lieutenant Keith Smith, who were the first to fly from England to Australia in response to an Australian Government prize of 10,000 Australian Pounds.

The Australian government's gamble to raise the profile of aviation paid off and the country went mad with excitement that the "mother country" was only 10 days away. The trip had taken 135 flying hours for the 12,651mile (21,001km) journey. Inspired by the feat, Queensland and Northern Territory Aerial Services (QANTAS) was launched to link outback Australian towns.

An English trade journal summed up the role of the pioneers, when it said, "One of the things that must be done is to create and maintain public interest and confidence in the aeroplane as a means of travel."

For the US public, that "someone" was the boyish Charles Lindbergh who, after his epic solo crossing of the Atlantic in 1927, was elevated to rock-star status. Tens of thousands hailed him when he landed in Paris but that was nothing to the adulation that swept America. Hundreds of thousands greeted him when he sailed into New York, while 30 million listened to him or saw him talk during a tour of 82 cities in 48 states. The US fell in love with Lindbergh and the aircraft.

In Australia, aviation fever struck after Bert Hinkler's solo flight from England to Australia in 1928. He was dubbed Australia's Lindbergh. But it was Sir Charles Kingsford Smith, who really fired Australia's passion for aviation when he crossed the Pacific from east to west in April 1928. More than a quarter of a million people welcomed "Smithy" to Sydney. He went on to pioneer many routes, including crossing the Tasman Sea between Australia and New Zealand in both directions. He was also the first to cross the Atlantic from east to west—against the prevailing winds.

THE ROLE OF AIRMAIL

In the US, the government supported air services with mail contracts and by the mid-1920s regular airmail services were widespread. While the value of aircraft moving mail was well accepted, the adoption of air transport for passengers was sluggish. However, Lindbergh's inspirational feat changed all that.

Dornier Do X. In 1929 it grabbed the headlines but its performance and economics were abysmal, despite 12 engines. On one test flight however it carried 169 passengers that included nine stowaways.
Wouter Sikkema

Americans embraced air travel on such aircraft as the new Boeing Model 80 trimotor, Fokker F.V11b and Ford-Stout 5-AT. Technology was starting to kick in. The Boeing Model 80 could carry 12 passengers and featured an enclosed cabin, forced-air heat, reading lights, leather upholstery and hot and cold running water. However, the airlines still needed the mail contracts to survive. No aircraft could yet make money just by hauling passengers.

While passengers were better off, the severity of vibration from the engines made long trips almost unbearable, and most aircrafts' seats were wicker affairs to save weight. On landing, mud sucked in by cabin vents splattered all over the passengers. For the privilege of enduring all that, the airfare for a transcontinental trip in the early 1930s was $258 - the equivalent of seven weeks' salary.

The Boeing 247 was a revolution in its day but not compared with today's 777. The 247 carried 10 passengers at 155mph (249km/h) and had a range of 485 miles(781km). The standard 777, shown here, can carry 375 passengers in two classes 4,568miles (7,352km) at 600m/hr (966km/h).
Boeing

America's new-found passion for travel and the advent of more powerful engines saw Boeing and its associated airline companies—the forerunners of today's United Airlines, develop the Boeing 247, which was the world's first truly modern airliner. The twin-engine 247 made its first flight in February 1933 and was a whopping 30 miles (48 km) per hour faster than the Ford Tri-motor - the best aircraft to that date. Nothing by today's standards but, in 1933, the 247 cut four hours off the US transcontinental flight making the crossing in just 19 hours.

But even greater performances were on the way. United's order for sixty 247s tied up Boeing's production lines forcing archrival Transcontinental and Western Airlines (TWA) to issue a request for the manufacture of an aircraft with tough specifications that would blitz the 247. Donald Douglas, who had founded the Douglas Aircraft Company in Santa Monica California, responded and set his engineers to work.

TWA had set Douglas a demanding challenge for that time of building an aircraft that could carry 10 passengers at a top speed of 185mph (300km/h) and a range of 1,080 miles (1,740km). The prototype twin-engine DC-1 which resulted was a stunning success and was able to meet all of TWA technical advisor, Lindbergh's very demanding specifications of being able to climb after take-off and maintain level flight on a single engine over any segment of the TWA route. In the test, the DC-1 exceeded the performance promised and was able to take off from Winslow, Arizona at its maximum take-off weight of 18,000lb (8,165 kgs), lose an engine on take-off and beat a Ford Tri-motor to Albuquerque, New Mexico by 15 minutes, despite the Tri-motor setting out 15 minutes before it. Needless to say, Douglas won the business but interestingly TWA's bankers baulked at lending the airline the necessary money to buy the aircraft because they didn't believe that such an aircraft could be built.

The DC-1 and its slightly bigger successor, the DC-2, which was put into full production, were crammed with new technological and safety features - variable speed propellers, auto-pilot, powered brakes, wing flaps, retractable landing gear systems and duplicate instruments in the cockpit. But these features came at a price causing design time to increase four-fold to 4,000 man-weeks over and above the1920s designs.

BRITISH DISBELIEF

Not only were TWA's bankers skeptical, so were the British. In 1934, when Sir Roy Fedden, head of Bristol Engines, showed a photo of the DC-2 to the UK Government Technical Directorate Development staff, they refused to believe it was a real aircraft. The British press ridiculed KLM-Royal Dutch Airline's entry of its shiny new DC-2 into the MacRoberston Air Race from London to Australia in 1934.

Douglas DC-2 used in the 50th year re-enactment of the MacRobertson Air Race of 1934, where a KLM DC-2 came second to an especially designed racing plane.

Incredibly clean lines of the Douglas DC-1. *TWA*

They were skeptical that the DC-2 could compete against the best from Europe. Newspapers described the American aircraft's entry as "American propaganda" and "an audacious assumption that such a ship could expect to compete with the fastest planes and designs from the continent."
But compete it did and despite carrying a full crew of five together with three passengers, traveling 1,000 miles (1,852 km) further and making five more stops to meet the obligations of its airline's regular Amsterdam-Batavia (Indonesia) service, the DC-2 came in second to a souped-up racing plane, the twin-engine de Havilland Comet, which carried just two pilots. As a result, airlines all over the world beat a path to the Douglas Aircraft Company and orders flowed in.

THE DC-3 - THE PLANE THAT CHANGED THE WORLD.

In response to a 1934 American Airlines' request for a version of the DC-2 which could accommodate beds, Douglas widened the fuselage and the Douglas Sleeper Transport (DST), later known as the DC-3, was born.

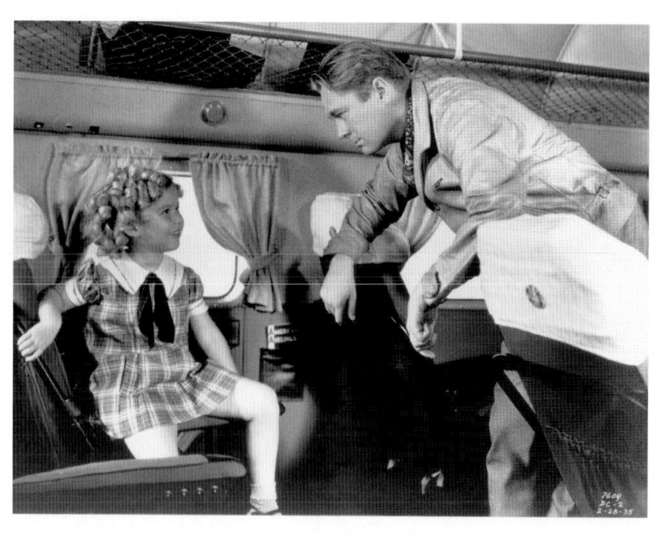

Movie stars such as Shirley Temple (seen here in a DC-2) promoted air travel. *Boeing Historical Archives*

During daytime flights, the DC-3 could carry 21 seated passengers and by virtue of better streamlining, it had only marginally more drag than the DC-2, despite carrying 50% more passengers. Compared to the Ford Tri-motor, which dominated US airlines in 1930, the DC-3 was a stunning advance in economics and technology.

Per passenger, the DC-3 was 40% more fuel efficient even though its speed had increased by almost 100%. And for the first time, speed had become an important factor in airline costs. With the DC-3, there was a massive decrease in crew and maintenance costs as a result of the greatly reduced flying time for a given sector. Passengers now viewed air travel as a pleasurable experience instead of an ordeal. Douglas added soundproofing, heating, fully upholstered chairs, galleys, reading lights, air vents and a big toilet.

Movie stars flocked to the Douglas Transport adding greater credibility for the public. Shirley Temple - everyone's darling in the 1930s - was seen singing on a DC-2. American Airlines was able to boast that one DC-3 flight's passenger list included Mary Pickford, Douglas Fairbanks Jr., Murray Silverstone, head of United Artists, Charles Chaplin and Sam Goldwyn. The dominance of the DC-3 in the US was extraordinary. In the 12 months from 1939-40, 80% of all flights were performed by DC-3s and the industry enjoyed a 100% safety record.

The DC-3 was sponsored by American Airlines. *Boeing Historical Archives*

AIRSHIPS

Without doubt the grandest flying machines were the monstrous airships. Disgracefully discharged from the army in 1887, Ferdinand Von Zeppelin, rather than disappear from the public eye, turned his considerable talents to flying and set about designing an airship.

Graf Zeppelin Airship touches down in England. The airship went around the world in 1929.

After several attempts he came up with the LZ2 which was 420 ft (128m) long and was powered by two 85hp engines which produced a leisurely speed of 50kmh. Hugh Eckner, a journalist at the Zeppelin's first flight, who was unable to muster any enthusiasm for the lumbering structure at the time, later joined the Zeppelin company and became the world's greatest airship captain.

Zeppelin's first airship flew in 1900 and between 1910 and 1914 his LZ7 carried 35,000 passengers between many cities in Germany, including Dusseldorf, Frankfurt and Baden-Baden. By the start of the WWI in 1914, although there were some safety incidents, there had been no injuries. However, after WWI, in a tragic episode, the British R38 crashed killing 44 people.

Despite this, the Germans went on to greater success and in 1929 the Graf Zeppelin flew around the world in 21 days carrying 20 passengers and a crew of 41. The Graf Zeppelin had 25 twin-berth cabins, a restaurant and lounge and linked Germany with South America and the US.

The biggest of all airships eventually brought the era to an end. The LZ129, christened the Hindenburg, began service in 1931. She was a magnificent airship with comforts never surpassed, even today. She could accommodate 50 passengers in comfort similar to a grand hotel. But the era of the airship came to a spectacular and tragic end at Lakehurst, New Jersey, on 6 May 1937, when the Hindenburg exploded in bad weather, killing 36. Incredibly, 62 passengers and crew survived. The cause of the explosion remains a mystery with sabotage or static discharge suspected.

BOATS THAT FLY

During these formative years, flying boats emerged as an alternative to ships for intercontinental travel from the late 1920s. The aircraft could land in calm sea at most remote destinations thus removing the need for expensive runways. The flying boats along with airships became the true ocean liners of the air up to the outbreak of WWII.

Golf putting practice on a Qantas Short S.32 C Class flying boat in 1938. *Qantas*

Pan American World Airways (Pan Am), by virtue of its globe-girdling network, was a leader in the development of the flying boat and overseas route networks. In 1935, it launched a trans-Pacific service with a Martin M-130 flying boat which took seven days to fly from San Francisco to the Philippines. But the challenge was enormous for the four-engine aircraft which could only carry a 2,000lb (907kg) payload (normally mail) to its maximum range of 2,760m (4,445km) - San Francisco to Honolulu.

Nonetheless, this was an incredible feat for the day. Its normal load of 32 passengers was carried in luxurious surroundings. The M-130 featured a host of technological advances, such as flush riveting affording a perfectly smooth skin, autopilot and constant speed propellers, which enabled automatic adjustment of propellers to suit speed and engine power in order to gain optimum economy.

Qantas and Imperial Airways (later BOAC/British Airways) were also at the forefront of operating flying boats, with the Shorts Empire flying boat operating sections of what was to become known as the "Kangaroo Route" from London to Sydney from 1937. The ultimate pre-war flying boat, however, was the Boeing 314 Clipper which

Short S.32 C Class flying boat. The S.32 formed the back bone of the mail and passenger flights between the UK and Australia. Cargo was also carried including the first seafood exports from Australia—three dozen oysters to Singapore, which built up to 400 dozen a week before the outbreak of war.

Imperial Airways HP-45 which could carry 12 passengers.
British Airways Museum

could carry an economic payload of 35 passengers across the Atlantic in either direction. Key to its performance were the aircraft's four 1,550hp Wright Cyclone engines - the most powerful engines fitted to any pre-war aircraft. By comparison, the DC-3 engines were 1,200hp.

The return fare across the Atlantic in those days was $675, equivalent to 20 weeks' salary at the time. The 314 had a 10-man flight crew and required a marathon inspection at the end of each flight. No less than 1,500 inspections had to be made by 200 men. The turnaround took six days when the service first started in 1939. This was later reduced to 48 hours. Today's 747 can be turned around in just 48 minutes.

The 314 was a magnificent aircraft by any standards. It could carry 74 passengers to 1,725 miles (2,778km) or 35 on the much longer trans-Atlantic trip. There was a 14-seat dining room and sleeping berths for all passengers.

Boeing 314 Clipper. The first aircraft to offer a regular passenger service across the North Atlantic.
Boeing Historical Archives

PISTON DEVELOPMENTS

Donald Douglas, who had shown the world how to fly with his DC-3 - and had also shown the airlines how to make money - said that the single greatest influence on aircraft design was the spectacular development of both piston and jet engines.

Douglas DC-4E - protoype of DC-4

As this book will explain, the rapid progress was to be the undoing of many aircraft designs. At the end of WWI, the most powerful engines available were water-cooled and developed up to 400hp but were unreliable and had to be overhauled every 75 hours of flying. There was also a glut of engines thus little further research and development was done on engines until the mid-1920s. By the late 1920s, engine development had resulted in air-cooled engines with the capacity of up to 500hp and reliability leapt.

Two of the most important contributions to the increase in power of the piston engine were the increase in the octane of fuel and propeller developments such as variable pitch. The improvements in fuel octane were staggering. In 1930 when 100 Octane fuel became available, the US engine maker Wright was able to increase its R-1820 Cyclone engine from 500 to 1,200hp.

With the variable pitch propeller developed by the British, the pilot could alter the propeller blades' angle for maximum efficiency for each phase of flight. On the 247 of 1933, because of the variable pitch propeller, Boeing was able to reduce the 247D's take-off run by 21%, increase the rate of climb by 22% and lift the cruising speed by nearly 6%.

By the end of the 1930s, piston engines had evolved from a post-war best of 400hp with an overhaul rate of 75 hours to 1,550hp engines with a six-fold improvement in overhaul rate. Of critical importance was the accompanying 50% decrease in engine weight per horsepower developed. More powerful engines also meant better speed. In the early 1920s, aircraft struggled to get above 100mph. Boeing's 80A trimotor could top 125mph (231km/h) in the late 1920s but that quickly gave way to the 247 and DC-2/3 that reached 196mph (363km/h) by 1934. When war broke out, some European designs such as the German Focke Wulf Condor could reach 225mph (417km/h).
The Condor set a spectacular record in 1938 when it flew non-stop from Berlin to New York in 24 hours.

FARES TUMBLE

These significant engine advances translated into much lower fares for passengers. In 1919, fledgling airlines charged $85 for the 90-minute trip from Miami to Nassau, which was similar to the rate charged between London and Paris.

Imperial Airways Armstrong Whitworth Ensign, which could carry up to 40 passengers, first flew in 1938.
British Airways Museum

This 185-mile journey cost 45 cents a mile. By 1930, just before the arrival of the 247 and the DC-2, fares were about 12 to 15 cents a mile on domestic US routes and 15 to 20 cents a mile on Pan Am's Latin American routes because of higher costs.

With the advent of the DC-2, fares plunged to six cents a mile on US flights while Pan Am's Latin American flights averaged 10 cents a mile. By 1940, US domestic fares had dropped to an average of five cents a mile for mostly short-ranged flights. On longer-haul routes, passengers could fly from Port Washington, New York to Southampton, England for $375 one way or eleven weeks' salary.

THE JET ENGINE AND PRESSURIZATION GIVE GREAT PROMISE

Sir Frank Whittle as a student in 1928, wrote a thesis on jet propulsion which aroused little interest in the UK. Whittle secured a patent for the design of a jet engine in 1931 but it was not until 1935, when the patent had lapsed, that he was able to form Power Jets Limited to build a prototype.

Two years later, the first jet engine was bench-tested. Hans von Ohain, working with aircraft designer, Ernst Heinkel in Germany, test flew a jet engine that developed 838lb of thrust fitted to the He178 jet fighter on 27 August 1939. The aircraft flew at an astonishing speed of 435mph (806km/h). Whittle's engine, which produced 860lbs of thrust, flew in May 1941 in a Gloster E28 fighter and reached a speed of 338mph (561km/h). By comparison, today's biggest jet engine, the GE90-115B, develops 115,000lbs of thrust.

In the early 1930s, German and French manufacturers Junkers and Farman both demonstrated pressurized cabins for crews on high-altitude aircraft. In 1937, the US Army Air Corps, later the US Air Force, validated the Lockheed XC-35 with a pressure cabin. Boeing seized the concept and designed the Boeing 307, which was a four-engine aircraft that could carry 32 passengers and fly at altitudes of 20,000ft (6,096m) above most of the worst weather. TWA put the aircraft into service between Los Angeles and New York via Chicago in 1940.

After six years of restoration work by Boeing employees and retirees, an ex-Pan Am Boeing 307 "Clipper Flying Cloud" took to the air in 2001. Many Boeing suppliers donated parts and furnishings to bring the 307 back to her original glory. The aircraft is now displayed in the Steven Udvar-Hazy Centre at Washington Dulles International Airport, companion of the Smithsonian's National Air and Space Museum. *Joe Walker*

CHAPTER 2

A New Dawn

More than anything else the sensation is one of perfect peace mingled with an excitement that strains every nerve to the utmost, if you can conceive of such a combination.
— Wilbur Wright on his first flight December 17, 1903 —

AIRCRAFT DEVELOPMENT IN THE US

When war broke out in 1939, US-designed aircraft were at the forefront of reliability and economy. So much so that 95% of all passengers around the world flew on Douglas DC-2 or DC-3 aircraft. These extraordinary numbers were also a reflection of the affluence of the US and the rapid development of the US airline system.

BOAC Lancastrian, an interim passenger conversion of the famous Lancaster bomber. Six passengers sat sideways facing windows on one side. *Qantas*

Aircraft to replace the DC-3 were already taking to the air as war clouds gathered. The forerunner of the Douglas DC-4 Skymaster, the DC-4E, had flown in 1938 and the pressurized Boeing 307 entered airline service in early 1940. Lockheed's famous Kelly Johnson was designing the Constellation to meet a specification from billionaire recluse and aviation enthusiast, Howard Hughes, who now owned TWA. These aircraft were years ahead of anything that the Europeans were producing.

While the war halted airline development of the new Douglas and Lockheed designs, the US Army Air Corp ordered the designs as military transports, thus underwriting development costs and enabling the aircraft to be further developed. Interestingly, before the US entered the war, the US airlines had rejected the DC-4E which was fitted with a host of luxuries for passengers. It was thought to be uneconomical because the fittings added a great deal of weight causing it to be too costly. A much leaner DC-4 had been ordered by the airlines before Pearl Harbor as they focused on doing away with the frills to get fares down.

Before the outbreak of war, a number of German designs were the equal of American aircraft from a technical aspect but were not as efficient. British commercial aircraft were however, significantly inferior in economics and technology but were high on luxury by virtue of the demands of Imperial Airways for opulent travel for its passengers. During the war, Douglas built more than 1,000 DC-4/C-54 transports and gained invaluable experience in four-engine transports. A much more potent model, the DC-6 was developed in the later stages of the war to compete with Lockheed's Constellation, which was pressurized for much greater passenger comfort. The pressurized version of the DC-4 was funded by the US Government under the title of the "Skymaster Improvement Program."

The XC-112A or DC-6 flew in February 1946 and the first service with American Airlines occurred in April 1947. Lockheed's Constellation flew in 1943 under military guise and entered service with TWA early in 1946. While US airlines were delivered brand-new, state-of-the-art aircraft together with converted military DC-4s and Constellations, most countries like England had to make do with converted bombers, such as the Lancastrian-civil version of the famous Lancaster—which could haul just a handful of passengers in a cramped and noisy cabin. The Europeans of course had been devastated by war with their factories in ruins and only British factories were able to churn out bombers and fighters to ensure the country's survival.

Lockheed's giant double deck 180-passenger Constitution, built for the US Navy, was powered by four 3,500hp piston engines but needed six JATO rockets to help take-off performance. *Lockheed*

ENGINE POWER SURGES

The reliability and performance of piston engines soared through the war. In 1939, the Boeing 314 had the biggest engine in commercial service, developing a respectable 1,550hp. By war's end, the USAF B-36's Pratt & Whitney engines produced 3,500hp.

At the same time, the engine's fuel consumption dropped, its weight was reduced and the time between overhauls almost doubled to 1,000 hours. Without doubt though, the greatest improvement in performance was in the jet engine, which went from an unreliable novelty, producing about 850lb of thrust, to an awesome source of power. For example, engines such as the General Electric 1-40 produced 4,000lb of thrust for the Lockheed P-80 Shooting Star fighter in 1945.

andover of the first two DC-6s to American Airlines and United Airlines

Constellation cabin.
Qantas

An American Airlines advertisement extolling the virtues of the speed of the new DC-7

TECHNOLOGY LEAPS

Aircraft like the Constellation and DC-4 featured a host of innovations such as hydraulic power-boosted controls and thermal de-icing of the wing, tail-plane and fin leading edges. Additionally, the Constellation's sleek design translated into exceptional performance:

- Its top speed was 347mph (643 km/h), faster than any WWII bomber and as fast as most fighters.
- It met the TWA specification of one mile per gallon of fuel at the high cruising speed of 275mph (509km/h) at only 52.5% power.
- It exceeded its payload range by a wide margin.
- It was able to fly at 15,000ft (4,572m) with two of its four engines out.

Demonstrating that stunning performance were Howard Hughes and TWA's Chief Executive Jack Frye, who piloted the first production Constellation, or "Connie" as it became known, from Burbank in California to Washington DC in just 6 hours and 57 minutes with 14 passengers on board.

EXPERIENCE, EXPERIENCE, EXPERIENCE

Not only did American-built aircraft dominate post-war aviation but the country's airlines had amassed a vast amount of experience transporting troops and supplies around the US and overseas. Pan American was to the forefront of this effort and at the peak of the war had 88,000 employees. The airline's small fleet of twelve 314s clocked up a staggering 7.5 million miles (12.07 million kms) and carried 61,000 passengers. In all, Pan Am's entire fleet, which included 75 DC-3s, flew more than 90 million aircraft miles (144.9 million km) and 18,459 ocean crossings.

e only aircraft to challenge the dominance of US
craft was the Vickers Viscount shown here in the
lors of New Zealand's National Airways Corporation.
ve Dyball

TWA and American Airlines also flew across the Atlantic to England, Africa and Asia. United Airlines flew regular flights to Australia. Northwest flew to Alaska and Eastern Airlines went to Brazil. This vast experience was to form the basis for US domination of the airways for decades to come, and it would not be until the 1960s that European airlines would catch up. Asian airlines followed in the 1970s. From an aircraft perspective, US manufacturers would dominate until the advent of Airbus Industrie in 1970 (and it would take Airbus a further 32 years before drawing level with Boeing).

A Lockheed family get together in 1951. In the foreground is the Super Constellation, then an Air France 749, three pre-war aircraft the Model 10 (1934), Model 12 (1936) Model 18 (1939). The giant in the background is one of two XR60-1 Constitutions built for the US Navy. The aircraft could carry 180 and Pan Am was interested when the aircraft was first mooted in 1940. *Qantas*

Bristol Brabazon *BAE SYSTEMS*

SMASHED DREAMS - THE BRABAZON COMMITTEE

In 1942, the British Government commissioned Lord Brabazon, the first Briton to fly in 1908, to chair a committee to recommend post-war aircraft designs. The committee suggested a series of concepts but two, which were intended to give the lead to British industry for the trans-Atlantic run, were utter disasters.

The 8-engine Bristol Brabazon and the 10-engine Saunders Roe Princess flying boat were magnificent in concept but impractical. These designs epitomize the ideal of luxury travel for the privileged few. Some English manufacturers did not embrace the pre-war trend in the US towards increasing affordability of air travel, highlighted by the US airlines' rejection of the luxurious and overweight DC-4E. The eight-engine Brabazon had a dining salon, cocktail lounge and cinema, and all passengers had beds. Airline indifference, government bungling and cost overruns killed the project. However, British European Airlines (BEA) did show some interest in operating the aircraft in a 180-seat configuration. It would be wrong though to dismiss the work of the Brabazon committee because it led to the building of a number of successful projects, including the medium-range Vickers Viscount which achieved sales of 444 aircraft.

Bristol Brabazon on its maiden flight
BAE SYSTEMS

DOUGLAS VS LOCKHEED - PASSENGERS THE WINNER

In 1945, both Lockheed and Douglas, at the time the world's major civil aircraft manufacturers, had two awesome long-haul aircraft types in production - the Constellation and the DC-4/DC-6. The performance of these designs made every other US and European direct competitor irrelevant. Of course, the US manufacturers had the luxury of developing their designs through the war years, although in fact both aircraft types can trace their origins to the late 1930s.

United Airlines DC-6 *United Airlines*

For the first time, long-haul air travel had become easy, comfortable and more affordable, as Douglas and Lockheed improved the performance of their designs to gain an edge. The two manufacturers relentlessly sought engine performance improvements to wrest the crown of "Queen of the Skies" from each other. From 1942, when the DC-4 first flew, to the final development of the family of aircraft known as the DC-7C in 1955, Douglas was able to double the payload, take-off weight and range as well as increase the speed by more than 50%.

Lockheed's famous and graceful Constellation series boasted similar improvements. The major key to the advance in performance was engine development that saw an increase in horse power from 1,450hp on the DC-4 to 3,400hp on the DC-7C. In fact, the final development of the Constellation, the 1649 Starliner, bettered the DC-7C but arrived too late and was quickly overtaken by the arrival of jets.

Both manufacturers were able to make significant improvements to passenger comfort by moving engines further outboard, reducing cabin vibration and noise.
Another innovation was the use of bigger propellers with lower tip speeds, additional insulation and synchro-phasing of the propellers to reduce noise. Cabins featured soothing colors, wall murals, reading lights, air vents and soft curtains.
Another important addition through the 1950s was weather radar. The massive increase in range of the Douglas and Lockheed designs was to have a major impact on the flexibility and economics of airlines. For the first time on long-haul routes, airlines such as British Airways, Pan Am, SAS, TWA and Qantas were able to offer passengers non-stop flights between key cities across the oceans, which further cut fares by eliminating costly refueling stops. Douglas dubbed its DC-7C, the "Seven Seas" in recognition of its long-range performance. Polar flights and non-stop flights across the North Atlantic in all weather became commonplace.

PAN AM FIGHTS FOR TOURIST CLASS

The widespread introduction of four-engine land planes at the end of the war meant a substantial decline in costs for airlines. Pan Am was able to bring fares down 30% by 1950, while cargo rates had dropped 50%. But incredible as it may sound in today's environment, reducing fares was not an easy matter on some international routes for Pan Am's chief executive and founder Juan Trippe.

Boeing 377 passing over the Queen Elizabeth.

Trippe launched one of the first real no-frills services known as, "tourist class" between New York and San Juan in September 1948. The airline used DC-4s in a five-across arrangement, adding 19 seats to increase capacity to 63. The cabin crew was reduced to one and only soft drinks were served, while boxed dinners could be purchased before departure. The fare was $75 one-way compared to the normal $133 and within five months of the introduction of the service, passenger numbers had trebled. These services were extended to most South American destinations, with governments keen to make travel more affordable. By 1951, tourist class flights accounted for 20% of air travel.

But on routes to Europe, Trippe needed to get the approval of the airline cartel, the International Air Transport Association (IATA), plus a host of governments that controlled the major European airlines. This proved to be a marathon effort that took four years. But on 1 May 1952, a DC-6B Clipper Liberty Bell operated the first tourist class (similar

The turbine-powered Bristol Britannia conceived in 1947 for BOAC, had considerable teething problems with its engines that delayed its entry into service till early 1957 and thus its impact was greatly reduced as jets took over just two years later.
British Airways Museum

to economy) flights between New York and London. The one-way fare had been set at $270 compared with $395 for first class. The lower fare was achieved by upping the seating from 52 to 82 and the tourist section was five across rather than four. But tourist class passengers still retained the generous 40-inch (101cm) seat pitch, compared to today's standard of 32-inches (81.28cm).

In a publicity demonstration of how smooth the ride was in the Comet, passengers build a house of cards. *BAE SYSTEMS*

Tourist class was an instant hit. Traffic doubled within a year and the service was extended to Paris, Rome, Brussels, Frankfurt, Amsterdam and Glasgow. By 1954, tourist class was available on all Pan Am routes and most routes around the world. The effect of the fares was stunning, with traffic increasing by 37% in 1955 on the North Atlantic. In that year, (system-wide) 62% of passengers were traveling on tourist tickets.

Comet 1s: The first three BOAC Comet 1s in formation. The aircraft nearest the camera operated the first commercial jet service on 2 May 1952 from London to Johannesburg. *BAE SYSTEMS*

BOEING SOARS ON A GAMBLE

In the 20 or so years since Boeing introduced the 247, the company's commercial designs - the 307 Stratoliner, 314 Clipper and 377 Stratocruiser - had been blitzed by the war or more economical competitors.

The Dash-80 rolled out on 14 May, 1954
Boeing Historical Archives

In 1952, Boeing's chief William (Bill) Allen elected to take a massive gamble and in May, Boeing's board committed to building a jet transport prototype that would double as a military tanker and civil transport. Both Douglas and Lockheed had looked at jet-powered designs in the early 1950s but the reliability, noise and fuel consumption of jet engines troubled designers, besides their airline customers had made enormous investments in piston-engine models.

The British led the way with the ill-fated Comet 1 (which suffered three tragic crashes related to structural failures) and the turbo-prop Viscounts, the latter being the British industry's most successful design. By the mid-1950s, the promise of the jet was irresistible. Boeing's 707 would blitz all other aircraft then in service. Its performance was a quantum leap, offering almost double the speed and payload for 22% less cost per seat than the best piston-engine aircraft of the day - the DC-6B.

The new jets, such as the 707, had price tags double that of the piston engine aircraft they would replace, but produced three times the revenue. However, the gamble for Boeing almost backfired because its archrival Douglas, was building a more roomy aircraft the DC-8 with bigger engines. Pan Am's Trippe had convinced Pratt & Whitney to build a bigger jet engine, the J75 and ordered 25 DC-8s from Douglas powered by the engine. On the same day, he purchased 20 of the smaller 707-120s powered by the J57. United Airlines followed Pan Am's lead and ordered 30 DC-8s, because Boeing was reluctant to widen the 707 to accommodate six across seating. Boeing's Allen was forced to relent. A wider and longer 707, the -320, powered by the bigger engine was launched. From that point, Boeing's domination of the jet era was assured.

Qantas' first turbofan-powered 707-138B
Boeing

AIR TRAVEL SURPASSES SEA, BUS AND TRAIN

When the 707 entered service, there was a scramble to get a seat with the airline's flights running at an extraordinary 90.8% load factor. Passengers loved the jets. In the first five years of jet operations, Pan Am's overseas traffic doubled as the airline took delivery of more than 50 new jets.

Lockheed opted to try and capture the medium haul routes with its 100-seat turbine powered Electra, rather than a pure jet option. *Ansett Australia*

In 1960, for the first time, the number of passengers crossing the North Atlantic by air surpassed those on ocean liners. Pan Am was offering a return trip across the North Atlantic from only $298 - just slightly above what the one-way fare had been in 1952 - or just three weeks' average salary. For a flight from the west coast of the US to London, the return fare was $581, while Paris to Washington was $416 return. And in 1960, tourist class or economy class as it was now called, dominated air travel to the point where first class had been reduced to just a small section of the jet's cabin.

Not only were fares plummeting, so were traveling times. Australia's Qantas was able to slash the London-Sydney route from 48 to 27 hours, while the Sydney-San Francisco route tumbled from 27 hours to 18 hours. Interestingly, because of the enormous distances traveled by its passengers, Qantas had one of the biggest first class markets with 23% of its travelers opting for the front end in the mid-1960s.

In the US in 1950, air travel accounted for just 14% of travel, with bus and train accounting for 38% and 48% respectively. Thanks to the DC-6 and Constellation, by 1959 when the first US jets entered service 47% of travelers were in the air, while bus and train had been relegated to 24% and 29%. In the 10 years to 1959, air travel had leapt 250% in the US. Ten years later, the number of passengers traveling by air in the US would triple.

Douglas DC-8 made it first appearance on 9 April, 1958.

GAS-GUZZLING MADNESS

Cathay Pacific Convair 880M

No story of the progress of technology in aviation is complete without reference to some of the spectacular failures. Most notable was San Diego-based Convair, a division of General Dynamics. Its attempt to enter the jet age with the four-engine CV880 and CV990 as a follow-on to its successful twin-engine piston CV240, CV340 and CV440 family was a disaster.

The project was bedeviled by its sponsor Howard Hughes who wanted a smaller jet than the 707 for medium-haul routes for TWA. The story, or rather the fiasco that followed, saw $425million written off in 1961 - the biggest corporate loss in history, according to the January 1962 issue of Fortune magazine. Not only did Hughes demand a host of design changes, he even insisted on the aircraft having a metal skin with gold anodizing. At one stage, the aircraft was dubbed the Golden Arrow because of this. Later, the anodizing was dropped because of batch match problems and the name was changed to 880, denoting the speed it would travel in feet per second.

But competition was intense from Boeing, which reduced the size of the 707 to compete on the medium-haul routes and the resulting 720 won orders from many big airlines, including the giant United Airlines, after Convair refused to increase the width of its five-across jet. That loss led to drastic measures at Convair, which pursued American Airlines' business by guaranteeing an enhanced 880 would beat the competition coast-to-coast by 45 minutes.

American's chief C.R. Smith was taken by the speed. He wanted a faster jet to "beat the pants" off Texas-based Braniff Airlines, which had ordered the domestic 707 with the bigger intercontinental engines for its flights. Braniff marketed it as "The Jet with The Big Engines". American's plan was to use the Convair jet for the first class only "Blue Streak" service and configure American's 707s for all economy. American purchased 25 of what was essentially a new aircraft from Convair but the aircraft's radical changes were not made clear to the General Dynamic's board.

Problems with the aircraft, now called the 990, surfaced almost immediately on test flights and the aircraft was unable to meet its speed and fuel burn guarantees despite a host of modifications. When the production run finished with just 102 880s and 990s built, GD had written off $4.16 million per plane - more than they sold for. Any possibility of further sales of the Convair jets was dashed when Boeing announced its three-engine 727 tri-jet in 1960. Once again the advancement in size and reliability of engines had overtaken aircraft designers.

CHAPTER 3

The Folly of Supersonic Dreams

Its operation in a world beset by
fuel and energy crises makes no sense at all.
— Senator Cranston of California, regards the Concorde, 1974. —

FRENCH AND BRITISH LEAD

After seeing the Americans dominate the jet race with the 707 and DC-8, the Europeans viewed supersonic travel as the next logical step. This was the age of "speed and sputnik" and it seemed there were no limits to what could be achieved. In fact, studies for a British design for a supersonic transport can be traced way back to 1954, whereas evaluations were not undertaken in the US and France until 1956.

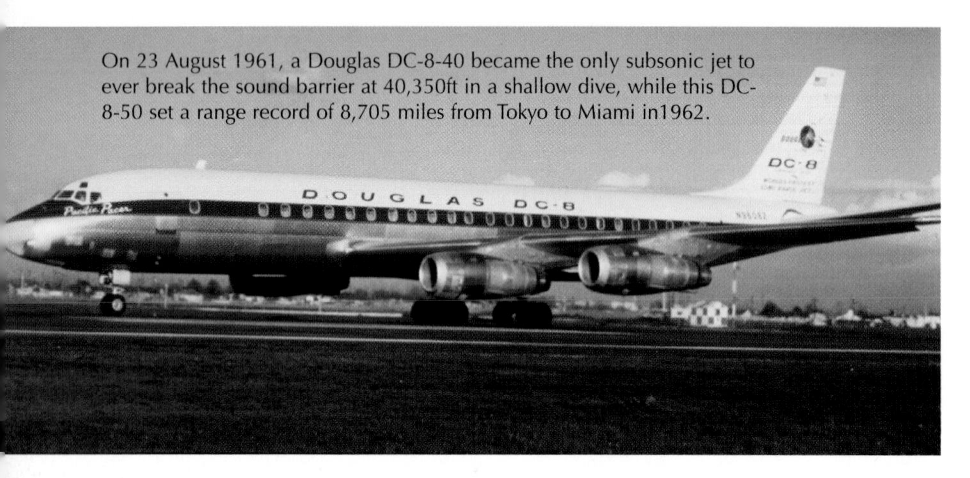

On 23 August 1961, a Douglas DC-8-40 became the only subsonic jet to ever break the sound barrier at 40,350ft in a shallow dive, while this DC-8-50 set a range record of 8,705 miles from Tokyo to Miami in1962.

The fact was that passengers were in love with speed, making this an effective marketing tool. Hollywood movies, such as Boeing Boeing, highlighted the problems that the new jet speed posed for two playboys dating a number of hostesses.

As we saw in the last chapter, US airlines and manufacturers were prepared to commit to new designs to gain 45 minutes on a five-hour trip. At the Paris Air Show in 1961, Sud Dassault proudly displayed a model of the supersonic Super Caravelle, which could carry 70 passengers to a range of 2,300 miles (3,703km). This served to galvanize British efforts and by March 1962 the French and British governments intervened to merge their efforts towards a joint Supersonic Transport (SST) which was signed off in November 1962.

But the agreement contained no provision for a ceiling on the costs (estimated to be EP170 million) nor was there any review or cancellation process agreed. There were already voices of caution. Pan Am's long-time adviser, the famous Charles Lindbergh, warned of the environmental and economic problems the aircraft would face. But Britain saw the SST as its entry ticket to the Common Market.

These hopes were dashed some months later, when President de Gaulle delivered his famous "No". However, as bait to the British, de Gaulle said that Britain's commitment to Europe would be judged by its progress on the SST which he referred to as Concorde - the first use of the name in public. Britain was committed to the program for the wrong reasons and the dye was cast for a disaster. The marriage between French and British manufacturers was also encountering problems as the Concorde's design was being thrashed out.

Britain tried to bale out in 1964, when the Labor Party was swept to power, but the French simply pointed out the massive penalties in damages, and new Aviation Minister Roy Jenkins who was trying to kill the project was shot down in flames. At different times, both France and Britain wanted to abandon the project but because of the wording of the treaty, the other would seize the opportunity to recoup its own expenditure and, of course, maintain that it wanted Concorde to carry on.

The aircraft in its original form was designed to carry up to 150 passengers 4,780 miles (7,696km) at Mach 2.2 just over twice the speed of sound. But it was discovered that the plane would fall short of New York by almost 434 miles (700km), so the aircraft was redesigned. As work

British Airways Concorde
British Airways

progressed, the very fine margins of the original concept became all too clear. Eventually, Concorde's payload (passengers and baggage) performance settled at about 100 passengers between London and New York.

It was not until 1965 that both sides finally agreed on the design and construction was started. The first Concorde to fly took off in March 1969 and by that stage options had been received for 80 Concordes from 18 airlines. But there were derisive howls about the development costs which were soaring faster than the aircraft. According to one estimate from the London-based Independent newspaper, the total write-off for the British and French governments was $23.4 billion in today's dollars, which works out to a taxpayer subsidy of a staggering $5,500 for every passenger who has flown on Concorde.

In early 1973, the world's airlines failed to convert their options to firm orders. Only government-owned Air France and BOAC put Concorde into service at no capital cost. A mere 16 production and four prototypes were built, with 13 introduced into service. The first commercial journey was not made until 21 January 1976, when a British Airways Concorde flew from Heathrow to Bahrain, at the same time as an Air France jet flew to Rio de Janeiro.

The transport of choice for the rich and famous, the Concorde was part of a world beyond the reach of most people. This made entering it seem all the more desirable and a 2003 BBC poll in the UK found that flying on Concorde was one of the top five things people wanted to do before they died. The problem for British Airways, according to its former Australian-born chief executive Rod Eddington, was that the number of high-fliers, particularly business types, is steadily dwindling.

RUSSIANS JOIN THE RACE

As the British and French efforts gained momentum, the Russians were also focused on supersonic travel - not to take the rich and famous across the Atlantic but to save time for scientists and military personnel. Government agencies believed that a fleet of 75 such aircraft was needed to cross the country and the Tupolev design bureau was tasked with the design in July 1963.

TU-144
NASA

The problem of the sonic boom was dismissed as being just like thunder - and possibly not nearly as bad as the wrath that would follow any complaints to the authorities. With its similar mission, it was inevitable that the Russian aircraft would be somewhat similar to the Concorde, although the arrest of Aeroflot's Paris manager at a cafe carrying a briefcase of Concorde plans gave rise to the "Concordski" label.

Nevertheless, the TU-144 was bigger, faster, more powerful and, many say, more advanced than the Concorde. Major differences between the two designs were the broad use of complex but lightweight titanium in the TU-144's hottest structures and the use of turbofan engines with afterburners.

The first aircraft was rolled out in 1968 and flew on 31 December of that year. However, when the first production aircraft appeared in 1972 it had undergone a complete re-design to address performance shortfalls. The aircraft was 20ft (7.93 mtrs) longer, had a new wing, its engines were repositioned and the aircraft could carry a much greater fuel load. Tragically, the second production aircraft crashed at the Paris Air Show in 1973, apparently after a cameraman standing in the cockpit fell on the controls when the pilot put the aircraft into a sharp turn to avoid another jet. The TU-144 also suffered from excessive vibration and fuel-burn, and ultra-high cabin noise.

Despite these problems and the air-show crash, the TU-144 went into service carrying cargo in 1975 followed by passenger flights in 1977. But passengers hated the cramped five-across cabin (Concorde is four across) and the vibration and noise sent passenger loads plummeting. After just 102 flights, the service of the TU-144 was terminated when a new, longer-ranged model, the TU-144D, crashed on a test flight, with some being scrapped and others stored. In 1994, a requirement by Boeing, McDonnell Douglas and NASA for research into high-speed transport saw one TU-144 restored to flying condition.

US SUPERSONIC STUMBLES ON REALITY

After a number of years of informal work, Boeing established a supersonic project office in January 1958 but because of the enormous cost of the project it was thought to be beyond the fiscal capability of any private company. Despite doubts and possibly because Pan Am's order for six Concordes in 1963 forced the issue, President John Kennedy launched the US counter to the Concorde.

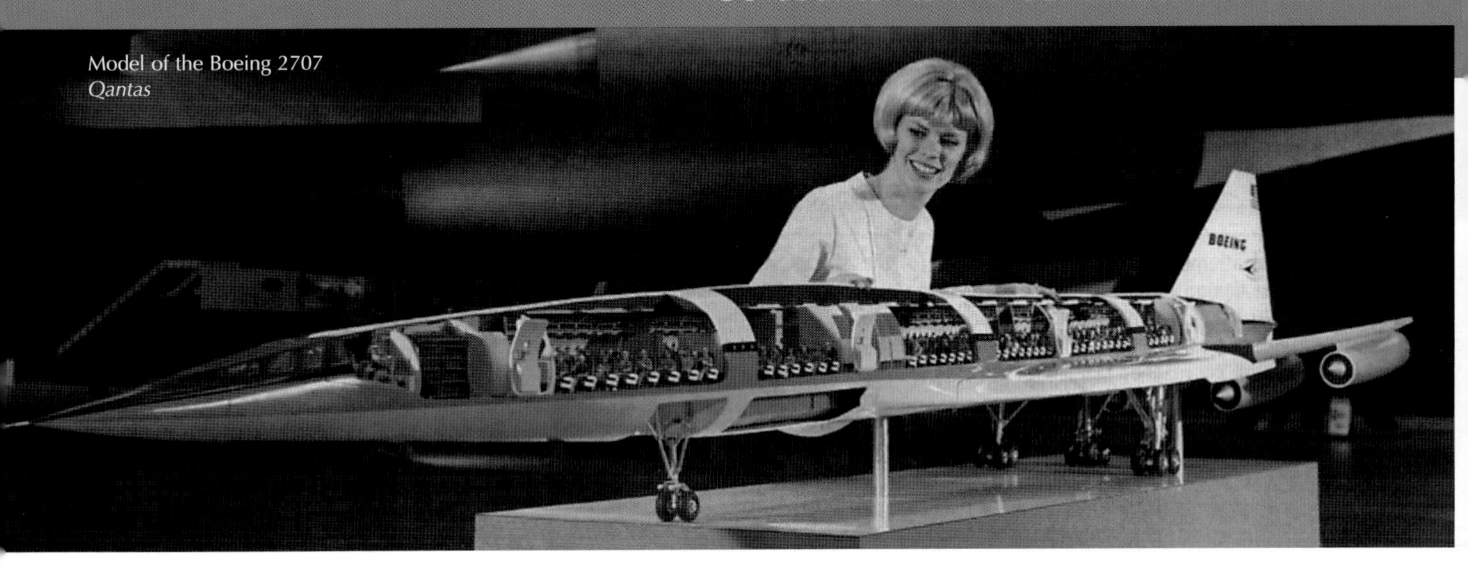

Model of the Boeing 2707
Qantas

The US Federal Aviation Authority (FAA) called for tenders from the US industry to build an aircraft superior to the Concorde. Douglas dropped out immediately, leaving Boeing, Lockheed and North American (which was building the supersonic bomber, the XB-70) to vie for the best design.

Boeing took the high-risk technology road with a radical swing-wing submission in contrast to the much more conservative delta-wing concept featured on the Concorde. Boeing was able to predict boldly that its SST design would have seat mile costs below those of the 707 and its design, along with that of Lockheed's, was selected for further development.

In December 1966, the US Government selected the Boeing design - the 2707-200. Boeing was to build two prototypes over four years at a cost of $1.44 billion in 1967 dollars. Twenty-six airlines took 122 delivery positions, with Pan Am, TWA and Alitalia to be the first to fly the jet.
The Boeing design was an extraordinary gamble based almost solely on the American "can do" attitude. Because the 2707 was to fly at Mach 2.7 (1,800mph or 2,898km/h) it would have to be built entirely from stainless steel and titanium, materials that were known at the time to be difficult to work. But worse was on the horizon.

As work progressed, it became evident that the swing-wing design was flawed and the 2707 would not be able to carry any passengers, while the level of noise it produced, together with the sonic boom, were all looming as insurmountable difficulties. Boeing scrapped the swing-wing in October 1968 in favor of a fixed delta-wing design but as construction proceeded with the mock-up, the aircraft ran into another obstacle much greater than the sound barrier - the cost hurdle. Congress was now deeply worried at both the rising cost of the 2707 project, which was obviously in disarray and growing public disquiet over its engine noise and sonic boom.

Finally, the union of the environmental lobby and the critics of government spending won the day and on 24 May 1971, the US Senate and the House of Representatives voted to kill the project. At the time, the US investment in the 2707 was $1 billion (1971 dollars) and

10 million man hours of work, and all the US had to show for it was a mountain of paperwork and a mock-up.

The 2707 was a watershed in commercial aviation. It was the first time that environmental concerns sounded the death knell of a project. At the same time, serious airline concerns over operating costs of aircraft had also become significant, flagging that the future in aviation was more than ever to be focused on reducing the cost of flying. Interestingly, just three years later, the Oil Crisis saw oil prices leap tenfold. The reality of supersonic travel had a host of problems.

- The Concorde and TU-144 were fuel guzzlers, burning one ton of fuel for every passenger carried across the North Atlantic.
- The Concorde fare across the North Atlantic was as much as 30 times higher than the cheapest discount economy fare.
- None of the combined 37 supersonic transport built was ever purchased by an airline in real terms.
- All had massive development costs.
- All had enormous environmental problems, with anyone standing half-a-mile from the Concorde "feeling" the sound let alone hearing it.

Time lapse photo of Boeing 2707 showing swing wing

- The sonic boom remains an insurmountable problem with no country allowing over-flights.

Boeing 2707 mock-up.
Qantas

MARKET DECEPTION

In an article in Airways (September 1995), noted industry historian R.E.G. Davies pointed out that very little research was ever done into the market for SSTs.

According to Davies, British Aircraft Corporation produced a brochure in 1970 claiming that more than 30% of the market would pay first class fares to travel Concorde. He noted that the reality was only 7% of the market was first class and it was declining. Today, it is less than 2% with many airlines dropping first class altogether on the North Atlantic. Davies also pointed out that on trans-Pacific flights an SST would be a scheduling nightmare because while the SST may take-off at the right time in New York it wouldn't arrive in Tokyo at an acceptable time because of its speed.

Concorde cockpit
British Airways

British Airways Concorde
British Airways

One of the many planned Concorde successors.
But like all others it remains just a dream.

MORE SUPERSONIC DREAMS

Despite the host of problems associated with supersonic travel, serious studies continue as aircraft manufacturers search for answers on how to break the environmental and economic barriers of supersonic flight. These barriers are proving vastly more difficult to break than the sound barrier.

During the mid-1990s, Boeing teamed with McDonnell Douglas, Pratt & Whitney and General Electric in a $1.3 billion NASA-funded study into a High Speed Civil Transport (HSCT). The study aircraft would carry 290 passengers at Mach 2.4 (1,585mph or 2,551km/h) and would be able to fly from Los Angeles to Tokyo in 4.3 hours compared to the typical 10.3 hours.

By 1996, Boeing, McDonnell Douglas and NASA defined a new baseline design for the HSCT program, called the Technology Concept Airplane (TCA) aimed at reducing the technical and economic risks. The TCA was slightly longer than the HSCT and featured a wing with greater sweep-back for lower cruise drag. The TCA was intended to be profitable at fares only marginally higher than tickets on jumbo jets and to go into service in 2015.

But the challenge of designing an aircraft that would carry three times the payload of the Concorde, double the distance, without hiking standard ticket prices proved impossible and the project was dropped. According to former Vice Chairman of Boeing and then CEO of McDonnell Douglas, Harry Stonecipher, designers of the TCA realized that the program did not make economic sense. "It was a gee-whiz aircraft that engineers love - but you have to get people to pay to fly on it."

One concept from McDonnell Douglas for a Concorde successor.

British Aerospace, now BAE SYSTEMS, also worked on a successor to its Concorde.

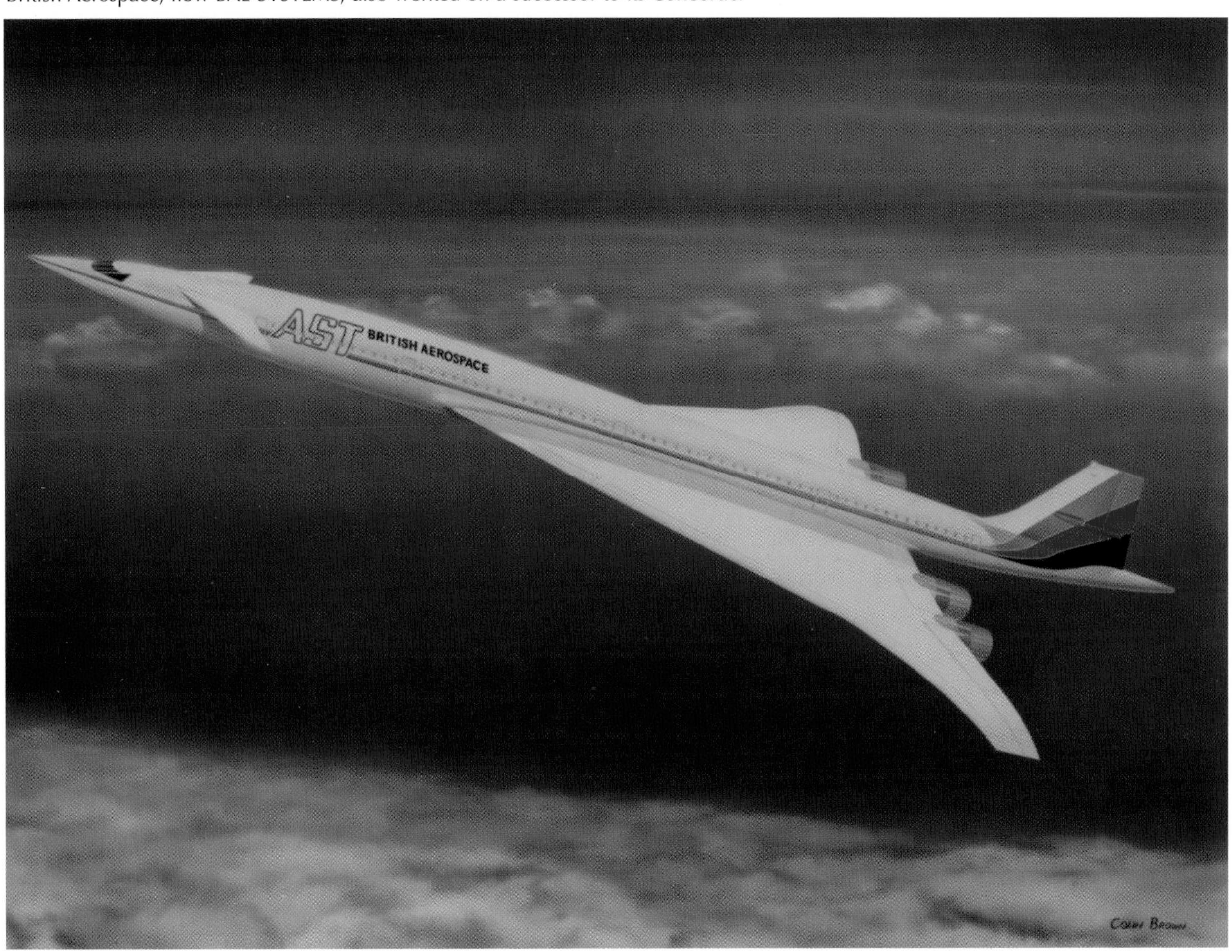

The three Concordes- looking like pterodactyls - taxi to British Airway's Engineering base. *Westralian Airports Corporation*

CONCORDE'S FINAL DAY

Hundreds of thousands turned up to Heathrow airport on Friday October 24, 2003 for a tearful farewell to the Concorde. They didn't care that the time machine was a dinosaur of the era of "speed and sputnik," and the biggest financial loss in aviation history. To them Concorde was a symbol of British and French technology. On a magnificent autumn evening three Concordes touched down on their last scheduled passenger flights with a host of VIP guests, such as Joan Collins and Sir David Frost - regular Concorde passengers - and fare-paying passengers.

Last portrait of four Concordes together.
British Airways

CHAPTER 4

The Jumbo Era

Aviation is proof, that given the will, we have the capacity to achieve the impossible.
— Eddie Rickenbacker —

The giant 747 rolls out into the Seattle sun.

The big fan owes its origins to a USAF requirement in the early 1960s for an aircraft that could lift 60 tons. This meant more than doubling the capability of the then biggest aircraft in the USAF inventory, the C-141 Starlifter freighter, which was of a similar size to the 707. At the same time, General Electric (GE) and Pratt and Whitney (P&W) had started design studies for the next generation engine to power such an aircraft. GE was smarting from the lack of success of its excellent CJ805 engine which powered the CV800/990 and saw the big fan as a way to leapfrog P&W and, to a lesser extent, Rolls Royce.

In 1964, GE had a half-scale engine running and in August 1965 won the engine competition for what had become the USAF's C-5A competition with its TF39. The engine would produce 41,000lb of thrust -125% more than the turbofan engines powering the 707-320 - and burn 25% less fuel. This was 40% less fuel than the turbojet engines that powered the first 707s. That leap in performance was of the same magnitude as the 707 over the piston-engine DC-7 just a few years before.

The three giant US airframe builders, Boeing, Douglas and Lockheed, were locked in battle to win the USAF's C-5A requirement and had poured more than four years' work into winning the project. Lockheed eventually won the day in September 1965 with the lowest bid, a fact that was to haunt the company for years to come.

Boeing 747 roll-out
Boeing

WHO NEEDED A BUSINESS PLAN?

Pan Am's founder and Chairman Juan Trippe, who had led the airline world into pushing for advances in technology, immediately turned to Lockheed to get a commercial version of the C-5A. This approach was spurned by Lockheed which felt - and rightly so - that it had plenty on its plate just building the military version of the aircraft.

The father of the 747 Joe Sutter (right) poses with John Travolta, movie star, aviation enthusiast and Qantas' "Ambassador at Large" and former President and CEO of Boeing Commercial Airplanes, Alan Mulally.

Douglas was mired in production problems with its fast-selling Super DC-8 and DC-9, so Trippe turned to Boeing. On 22 December, 1965, just 12 weeks after losing the C-5A competition, Boeing's chairman Bill Allen and Trippe jointly signed a letter of intent to launch the 747 program. It is impossible to find anyone who recalls if there was a definitive business plan for the 747. But traffic was booming for the airline industry which had enjoyed growth of 15% a year through the early 1960s as passengers flocked to jet aircraft.

Trippe was a man on a mission. He wanted to make travel affordable for everyone and he believed that the 747 with the enormous new engines could do just that. Trippe had introduced economy class, embraced the jet era and wanted this as his retirement swansong. The promising 747 was expected to cut operating costs by a staggering 30% over the 707. There was no doubt that the airline needed a bigger aircraft but did it need one twice the size of the biggest jet flying at that time?

Regardless of the reticence of other airlines, on 13 April 1966 Pan Am ordered 25 747s. These aircraft could carry about 500 passengers in an all-economy layout at a production cost of $22 million each. Other airlines were horrified. The realization dawned that they would have to buy the big jet themselves to stay competitive. Boeing had to build it and P&W had to power it. The die was cast. Boeing and Pan Am had put their futures on the line.

Pan Am 747 poses and dwarfs a Pan Am 707.
Boeing Historical Archive

TRIPPE'S DREAM IS BOEING'S NIGHTMARE

It is hard to imagine today, with more than 1,300 747s sold, that the aircraft almost bankrupted Boeing, P&W and Pan Am. Orders for the big jet only trickled in initially. Japan Airlines and Lufthansa wanted 3 each, BOAC 6, American 10, TWA 12 and United 5. These numbers were a far cry from the orders for 30-plus from each airline that launched the jet era. However, sales were not the major problem.

The overriding challenge was weight - and lots of it. Problems with the 747's weight and the engine power required to remedy the situation were almost a disaster. The story provides a fascinating insight into the complexities and challenges of building an aircraft that pushes technology to the edge.

The 60s was a "can do" era. America was going to send a man to the moon. The US, Britain, France and Russia were locked in a reckless battle to build and sell supersonic aircraft. Anything seemed possible.

When the 747 was first envisaged in late December 1965, it was to weigh 550,000lb (250,000kg) at take-off. When the order was signed six months later, Pan Am and Boeing agreed that the take-off weight of the 747 would not exceed 655,000lb (297,000kg). P&W committed to supply an engine with 41,000lb of thrust and increased it to 44,000lb within three years of the first deliveries. But airline demands for additional features had 747 weights soaring. For example, Boeing put air-conditioning ducts into the space behind the cockpit. Trippe said "not a chance", he wanted lounges in that area and that added 4,500lb (2,045kg).

By June 1967, the 747's take-off weight was 682,000lb (310,000kg) and climbing. Pan Am and Boeing agreed that the 747's range would drop by 900 miles (1,449km) to 5,000 miles (8,050km) and the maximum altitude would drop 2,000ft (610mtrs) to 33,000ft (10,060mtrs). While the change in altitude may not seem a problem, it was a significant one for the airlines operating the 747. Their aim was to cruise 10% faster than other jets. The reduction in altitude meant that the 747 would not be able to fly above them and thus would have to fly at a lower speed, on busy air routes.

But Boeing's problems couldn't be solved that easily. The company realized that it needed to increase the 747's weight to 710,000lb (322,727kg) and thus needed P&W to deliver an engine at 43,500lb of thrust three years ahead of its guarantee. Pan Am and P&W actually wanted to delay the program by nine months but Boeing insisted on pushing ahead.

According to some insiders at Boeing, the company threatened to cancel the 747 program unless the engine maker agreed to the additional thrust. Boeing realized that they had serious competition

Lockheed performed a dedication of its L-1011 Tristar plant on 20 July 1970 and were able to show to the world's media a "completed" Tristar before the DC-10 was rolled out.

looming from Lockheed and Rolls-Royce, which had started production of the Tristar, and McDonnell Douglas and GE with their DC-10.

P&W finally agreed to Boeing's terms. The engine manufacturer recognized that a quick way to get the extra thrust was to run the engine at higher temperatures but this would mean the engine was less efficient and would burn more fuel. To rectify the fuel consumption problem, however, meant running the engine at even higher temperatures.

Another vexing problem occurred in flight testing when the engines themselves distorted. A solution in the form of stiffening was found but deliveries of the big jet were consequently delayed by several months. P&W worked on other answers to the fuel consumption problem and finally, within about six months of the 747 entering service, fuel consumption was deemed acceptable.

Despite the many problems encountered in its manufacture, the birth of the 747 was an amazing feat. Pan Am took delivery of its first aircraft just 3-and-a-half years after its order was placed and that included a 10-month flight-test program. The first 747 service was carried out by Pan Am on 22 January 1970 - although it was the third attempt because of engine problems.

Air travel for the masses had arrived, but serious problems still existed. The 747's engines were giving the airline significant problems and Pan Am threatened to withhold $5 million per aircraft until at least 20 modifications were made. The airline also considered grounding its 747 fleet.

A solution was eventually found. Boeing and P&W agreed to a penalty. But while litigation receded, many problems remained.

Pan Am inexplicably ordered another eight 747s, placing enormous pressure on the airline, while the world's biggest aircraft manufacturer was deeply in debt and banks were owed $1.2 billion - $18 billion in 2003 dollars. The result was savage, with Boeing laying-off 64,000 staff to survive. For three years not one order was placed for the 747 and production fell from seven a month to below two. In fact, in 1974 Boeing produced as many 707s (20) as it did 747s.

KOLK'S MACHINE IS DOUGLAS AND LOCKHEED'S DOWNFALL

When the Pan Am order for the 747 was written up in "Aviation Week" on 18 April 1966, tucked away among the pages of text covering the mammoth order was a small break-out box titled "Airbus Plan."

DC-10 roll out special guests: front row L-R. James McDonnell, Chairman of McDonnell Douglas, then Vice President Spiro Agnew, then Governor of California and later President of the US, Ronald Reagan and then Honorary Chairman of McDonnell Douglas, Donald Douglas Snr.

The first DC-10 rolls-out

The story related not to Airbus Industrie, which was still a dream in a few far-sighted European minds, but an aircraft specification for a twin-engine aircraft that would seat 250 passengers, with a range of 2,125 miles (3,422km). The specification issued by American Airlines was drawn up by its chief engineer, Frank Kolk, and sent to Boeing, Douglas and Lockheed.
While Boeing had its hands full, Douglas and Lockheed seized upon the opportunity with vigor, which inevitably developed into a savage battle. Both companies saw the big advances in engine technology as opening up new markets to replace Boeing's top-selling medium-haul 727.

By mid-1967, the twin-engine concept had evolved into a three-engine aircraft to satisfy the range requirements of TWA for a transcontinental aircraft. It also alleviated the fear by airlines of operating a big 250-seat jet with just two engines over the Rocky Mountains. It is important to note that in 1967 engine reliability was nowhere near what it is today. The big turbofans had yet to establish a track record and airlines were skeptical about the claims from the engine makers.

What unfolded next was a disaster for Douglas, Lockheed and the airlines because the market for such an aircraft was simply not big enough for two players. Over three months of intrigue and bungling by all involved, two identical designs were committed into production in early 1968, with three engine makers. Neither airframe maker (Douglas nor Lockheed) would ever make money on the aircraft. Both the Lockheed L-1011 Tristar and the McDonnell Douglas DC-10 were technological marvels for their day and were instrumental in bringing airfares down but neither company was able to properly develop a family of models, which would have reduced costs for the airlines.

Lockheed planned to build an extended-range model, an extended fuselage and a twin-engine version of the Tristar, dubbed the Twinstar. Likewise, Douglas had similar plans and was able to launch an extended-range model the -30 and the -40. But the DC-10 Twin and a variety of stretched versions never left the drawing board until the MD-11 appeared in the late 1980s. These models, particularly the Twinstar or DC-10 Twin, would have allowed airlines to buy aircraft which had a high degree of commonality - enabling them to save on training costs and spares.

In the end, airlines had to turn to the twin-engine A300 or 767. When Douglas was unable to launch a stretch version of the DC-10, a number of airlines such as Air New Zealand were forced to buy 747s. Lockheed eventually exited the commercial market, while McDonnell Douglas severely weakened was forced into a merger with Boeing years later.

EUROPE FINALLY UNITES TO DOMINATE

The undisputed winner in the battle of the mini-jumbos was Airbus Industrie. France's Airbus program started in 1966 with engineer Roger Beteille and a secretary. Beteille sought the Kolk specifications for the twin-engine Airbus which he believed would give his own ideas greater credibility.

On roll-out the A300B1 poses for a family photo with the Concorde. It was to be a great omen, for Airbus sales went supersonic over the next 30 years. *Airbus*

Europeans had been searching for ways to cooperate for years to combat the dominance of the US airline industry. Britain, France, Germany, Holland, Sweden and Italy had formidable aircraft design and engineering capability but unity was an overriding problem. Only the Vickers Viscount, Sud Aviation Caravelle and Fokker's F-27 Friendship had established themselves as truly successful programs. A number of others had been dismal failures costing millions. However, by the late 1960s the economic reality of cooperation was finally overcoming national pride.

In 1966, France, Germany and Britain agreed to cooperate on the Airbus project and by April of the next year funds were flowing to the project team. However, by 1969, Britain's Rolls-Royce, which was to have responsibility for the Airbus engine, was committed to the Lockheed Tristar and withdrew from the program. Germany and France proceeded as equal partners and GE won the right to power the aircraft after agreeing to give major work to French engine company Snecma.

The founding fathers of Airbus were Beteille, German aerospace manufacturing genius Felix Kracht, Aerospatiale chairman Henri Ziegler and German politician Franz-Josef Strauss. The twin-engine jumbo, which is the most economic use of the big fan engine, was born. Well, almost.

Something was missing. Britain's Hawker Siddeley, now part of British Aerospace, was to build the wing but there was no government funding when Britain withdrew. Designing and building the wing is critical and the British had extensive experience in wing design. The Germans knew that and a solution was found and the German Government funded Hawker Siddeley's participation.

First flight of the A300B1 on 28 October 1972
Airbus

But while technically Airbus had struck the right formula, sales did not flow immediately. The company had a host of barriers to overcome, including import duty into the US, lack of a track record, airlines bloated with 747s, DC-10s and Tristars, and the fuel crisis of 1973.

The A300, however, was the right aircraft for the airlines, burning 25% less fuel per passenger than the 727-200, which it was designed to replace. And as a bonus it could haul two-and-a-half times more cargo than a 727 and five times more than the DC-9. The A300's two GE CF6-50 engines could develop upwards of 49,000lb of thrust - technology triumphed again.

By the mid-1970s, Airbus was starting to be noticed and sales slowly built for the A300 and the smaller A310. In the 30 years since, the company has sold more than 4,000 jets with a variety of designs, has 50% of the world backlog and is now building the world's biggest commercial passenger aircraft, the A380. Not surprisingly, Airbus has built its success on pushing technology to its limits and as one Airbus executive quipped: "We not only push the technology envelope, we have licked it."

The giant American breakthrough for Airbus. Eastern Airline's Frank Borman (centre) shakes on a deal to acquire A300B4s with Airbus's Bernard Lathiere (right) and Roger Beteille. Note one A300 to left of picture, beyond the model, is already painted in Eastern colours. *Airbus*

JUMBO CARGO

When the jumbo era arrived in the early 1970s, airlines were able to offer shippers an enormous amount of space at competitive rates and importantly under-floor containerised loading, which revolutionized the airfreight industry overnight. The 747, DC-10, Tristar and A300 could all carry underfloor containers, called LD3s. Pure freighter versions of the 747 and DC-10 also sold quickly.

TWA Tristar

But the enormous belly cargo capability of the jumbo jets was also a major factor in bringing down the cost of airfares and in 1970 Airbus was able to tout that its A300 would virtually pay for its operating costs from the belly cargo it carried – providing it was at express cargo rates.

Today, airlines such as Cathay Pacific, derive 30% of their revenue from cargo – most of which is carried under-floor, while EVA Air in Taiwan gets over 40% of its revenue from cargo.

Douglas was able to launch two longer ranged versions of DC-10, the -30 (shown) and the -40.

CHAPTER 5

Jet Development

In the entire history of jet transport there has never been an accident as a result of multiple engine failure due to independent causes.
— Airbus, ETOPS-Twins Through Time, 1989 —

Pan American Boeing 337 Stratocruiser

SIMPLICITY

Simplicity of the jet engine is one of the keys to its success. The jet pulls in air using a compressor which is studded with numerous small blades, shaped like aero-foils. That air, mixed with fuel, is burnt in the combustion chamber and expands out past the turbine, thereby driving the compressor before being expelled out the jet pipe.

By comparison, the piston engine had, by the 1950s, become hideously complicated, with complex cooling systems and electrical equipment. In fact, as the power of the piston engine increased, the number of engine failures increased.
In 1936, the DC-3's 1,000hp piston engines had a failure rate of 0.10 per 1,000 engine hours but the much larger 3,400hp piston engines that powered aircraft such as the Lockheed Constellation and DC-7C in 1955 had an engine failure rate of 0.89 per 1,000 hours.

Even bigger piston engines that would develop up to 6,000hp were undergoing testing but they were far too heavy and becoming more and more complicated with the result that reliability would inevitably diminish.

The answer was the jet. Development of the jet engine accelerated dramatically during WWII. The British shared Sir Frank Whittle's technology with the US, enabling General Electric (GE) to build a jet engine for America's first jet fighter. At the end of the war, German research was unlocked and both GE and archrival Pratt & Whitney (P&W) added German expertise to engine designs.

In 1948, both Bristol in the UK and P&W developed the dual spool jet engine, which essentially combined two engines into one with two compressors. P&W's concept evolved into the J57engine, which entered service with the USAF in 1953 and was fitted to the first 707 and DC-8s. The J57 was typical of what is termed the first generation of jets, dubbed turbojets, where the entire thrust comes from the hot stream of exhaust.

The "first" jet airliner was the Lancastrian which had two Rolls-Royce Merlin piston engines removed and two Nene jets added. The aircraft was never certified.

Anywhere on earth in 36 hours

by COMET

(Regd. U.S. Pat. Off.)

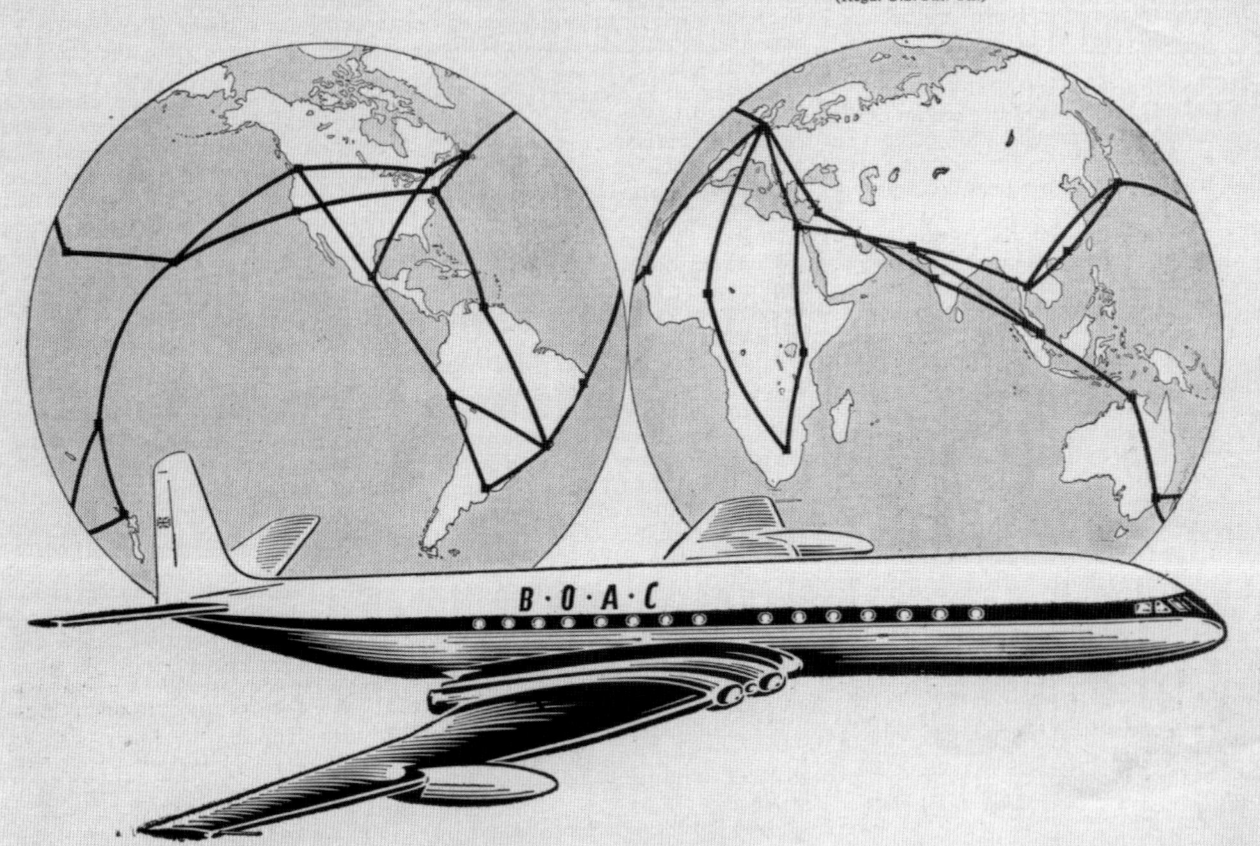

The new Comet, carrying 60 first-class passengers on 3,000-mile stages, is of ideal size for linking the world's many traffic centres. It makes possible high frequency with profit-earning loads.

The new Comet embodies all the experience gained in six years of pioneering jet airliner development. Behind it are 30,000 hours of airline service and half a million hours of operating experience with Rolls-Royce Avon engines.

B.O.A.C. will introduce the Comet 4 into world service in 1959.

DE HAVILLAND of GREAT BRITAIN

Engineers found that they could harness the enormous power of the jet exhaust by installing larger turbines to drive a large fan at the front of the engine and that this additional thrust could be then fed around the core of the engine. This development was called the bypass or turbofan and was first introduced in the early 1960s.

Frank Halford (de Havilland Engines), Sir Frank Whittle, Sir Geoffrey de Havilland and C. C. Walker (de Havilland Aircraft) pose with prototype Comet in 1949. *BAE SYSTEMS*

The turbofan not only produced more power but was also more economical. In addition, the slower air surrounding the hot jet had the effect of muffling engine's jet noise. The ratio of the bypass air to the core air or jet exhaust is called the bypass ratio (BPR).

The concept of the turbofan was originally patented in 1936 by Whittle. Tragically for England, just as few saw the value of the invention of the jet in 1929, little interest was shown in the turbofan. In a perplexing move, in 1943 the British Government told Whittle's company, Powerjets, to stop building jet engines just as Whittle was about to bench test a turbofan.

The BPR of the early turbofans, such as Pratt & Whitney's JT3D, was by today's standards a very modest 1.42 but, in 1960, it was a major breakthrough with airlines refitting their 707s and DC-8s with the more powerful engines.

In 1936, Whittle predicted that turbofan engines could have a bypass ratio of up to 60 but it would not be until 1965, when General Electric produced the TF39 for the giant military air-lifter, the C-5A Galaxy, that the large bypass engines would emerge.

The development of the turbofan engine led to larger aircraft such as the Super DC-8s that could seat up to 250. *McDonnell Douglas*

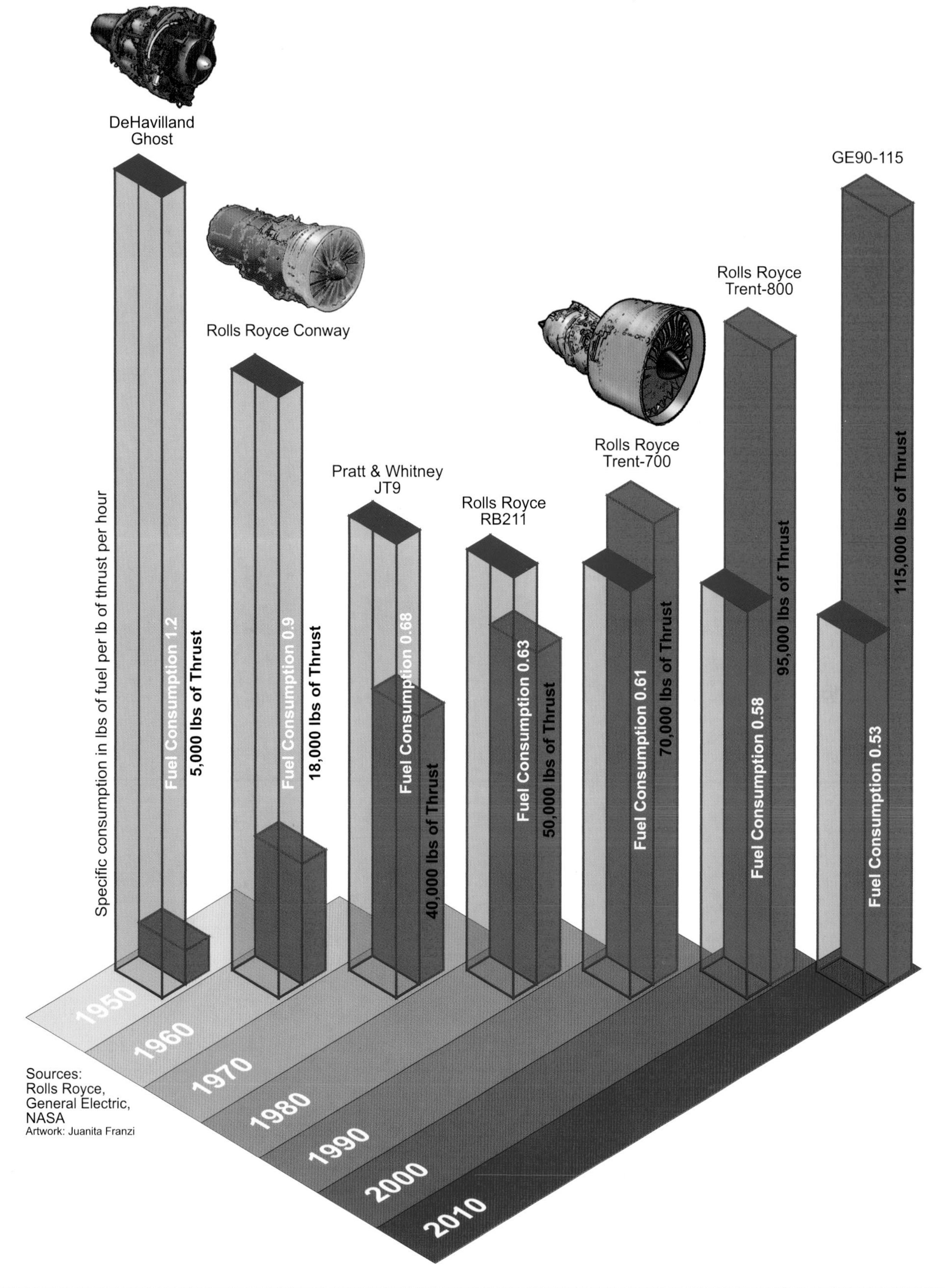

Power to Soar...
Development of the Jet Engine
Power and Efficiency
DeHavilland
Ghost
Rolls Royce Conway
Pratt & Whitney
JT9
Rolls Royce
RB211
Rolls Royce
Trent-700
Rolls Royce
Trent-800
GE90-115
Specific consumption in lbs of fuel per lb of thrust per hour
Fuel Consumption 1.2
5,000 lbs of Thrust
Fuel Consumption 0.9
18,000 lbs of Thrust
Fuel Consumption 0.68
40,000 lbs of Thrust
Fuel Consumption 0.63
50,000 lbs of Thrust
Fuel Consumption 0.61
70,000 lbs of Thrust
Fuel Consumption 0.58
95,000 lbs of Thrust
Fuel Consumption 0.53
115,000 lbs of Thrust
1950
1960
1970
1980
1990
2000
2010
Sources:
Rolls Royce,
General Electric,
NASA
Artwork: Juanita Franzi

THE BIG FAN

The TF39 and its civil version the CF6, Pratt & Whitney's JT9D and Rolls-Royce's RB211 all emerged in the late 1960s to power the first twin-aisle aircraft, the 747, DC-10, Tristar and A300.

Engines now had two quite distinct sections, the fan and the core. The core, where the combustion occurs, resembled a conventional turbojet seen on the first 707 but on the front was a huge, multi-bladed fan, which also provided a stunning backdrop for airline hostesses in promotional photographs.

All these engines produced about 40,000lb of thrust in the initial versions but it wasn't long before more demanding range/payload missions resulted in significant growth versions of more than 50,000lb.

In the case of the RB211, the engine thrust grew from the initial 40,000lb to more than 60,000lb and, compared to the engines that powered the Comet 1, thrust had grown 12 times while specific fuel consumption (sfc) had been halved.

General Electric's giant 115,000lb GE90-115B dwarfs co-author Christine Forbes Smith.

Hostesses squeeze into a Rolls-Royce Conway R.Co 12 engine - 17,500lbs of thrust. *Canadian Pacific Airlines*

THE GIANT FAN

But bigger jets were on the radar and engine makers such as Rolls-Royce were asked to produce engines with even more power for aircraft such as the A330, the MD-11 and later the 777. In most cases, an increase in the size of the core of the engine was required plus a new, bigger fan section. Rolls-Royce named its new engine the Trent 700/800 series, while P&W called its engine the PW4000, and GE launched the GE90. These engines or their derivatives have developed thrust of over 100,000lb, although not all are certified to that level.

To produce such engines, designers have had to balance the elements of weight, drag, noise and emissions. A major task is to improve fuel economy by improving its propulsive and thermal efficiency. Put simply, says Rolls-Royce, better efficiency means getting more power to propel the aircraft from the given amount of energy generated in the combustion process.

Thermal efficiency is improved by increasing the engine's overall pressure ratio and turbine entry temperature, and using better, more advanced components able to withstand higher temperatures.

An aircraft engine, unlike a power station turbine, calls for a unique set of parameters, namely very low weight and incredible reliability while at the same time producing high rotational speeds and coping with extremes of temperature.

An example of this conflict is the Turbine Entry Temperature (TET) at take-off, which is 1,600°C. Incredibly, this is about 300°C higher than the melting temperature of the material – nickel super alloys – from which the turbine blades are made. But the turbine blades don't melt due to internal cooling and the insulation of their surface by a film of air; so the blade surface does not experience the extreme temperature of the gas in which it operates. In a modern, high-pressure turbine, one four-inch-long aerofoil (blade) rotates so fast that it carries a centrifugal load similar to the weight of a London double-decker bus.

Any reduction in fuel consumption is, of course, a great saving in cost but more importantly it allows the airline to carry more payload (passengers and cargo) to earn more revenue. As the fuel load is typically two-thirds greater than the payload when full fuel load is carried, a 1% fuel saving can result in a 3% increase in payload.

The high bypass engines have had a dramatic effect on fuel consumption efficiency, according to P&W, which claims that in the 10 years from 1972 when jumbo jets entered widespread service, passenger numbers grew by 75% but total airline fuel consumption dropped slightly.

According to Rolls-Royce data, the overall fuel efficiency of commercial jet aircraft has improved by about 5% a year since the first jets took to the air. One of the other drivers of this has been the use of 3D aerodynamics in the turbine, giving even gas flow through each stage, increasing efficiency still further.

Other aerodynamic improvements include wide chord fan blades, an area in which Rolls-Royce lead the world, introducing the technology in 1984, over 10 years before its competitors. The blades currently used in the Trent 800 for the 777 are made with an advanced manufacturing process to produce a hollow construction which is strong, yet significantly lighter than earlier designs. Main impetus behind the aerodynamic improvements has been the developments in simulation software and computing power over the last 20 years. These advances have allowed accurate modelling of more and more complex situations deeper into the engine, pushing the component efficiencies to over 90%, giving a very fuel-efficient engine.

RELIABILITY

In fact, not only has engine fuel efficiency improved dramatically, so has the reliability. To illustrate, 20 years ago, the cost of the replacement parts and maintenance of the engine would equal its original purchase price within eight years. Today, that takes 30 years.

Boeing 727 roll-out 22 November 1962
Boeing Historical Archives

Convair 880 advert extolling the virtues of speed.

The reliability is reflected in the In-Flight Shut Down (IFSD) rate of jet engines, which has dropped from 0.9 shutdowns per 1,000 engine hours for the first turbojets to an incredible 0.005 shutdowns per 1,000 hours today. Boeing claims that today's turbofan engines are 50 times more reliable than the big piston engines and first turbojets. The extraordinary increase in reliability has had far-reaching effects on aircraft design, airline flexibility and costs, all contributing to a reduction in airfares.

Virtually all early jet-powered aircraft had four jet engines to enable them to cross the North Atlantic. The only significant exception was the twin-engine Sud-Est Caravelle designed for short ranges around Europe.

Another significant factor was the US Federal Aviation Administration's (FAA) Part 121 rule that prohibited two- and three-engine aircraft from flying further than 60 minutes from the nearest suitable airport. This was known as the 60-minute rule and was introduced in response to the limited reliability of piston engines.

The Boeing 767 (shown) and the A300 pioneered ETOPS. *Boeing*

RELIABILITY=ETOPS

The dramatic improvement in jet-engine reliability in the early 1960s prompted the FAA to exempt three-engine jets such as the 727 from the 60-minute rule.

By the late 1970s, jets were exhibiting a 10-fold improvement in reliability over piston-engine aircraft. Thus the FAA moved to ease the 60-minute rule to 120 minutes for twin-engine aircraft. This was introduced in 1985 under a program called Extended-Range Twin Engine Operations (ETOPS).

To get ETOPS approval, an airline had to demonstrate 12 months of satisfactory operation to agreed standards with an aircraft/engine combination that had an IFSD rate of 0.05/1,000 hours or better.

But it wasn't only engines that had to meet strict standards. ETOPS-certified aircraft had to have a stringent number of other elements including:

- Additional cooling monitoring
- Extra back-up hydraulic motor/generator
- A revised Engine Instrument Crew Alerting System (EICAS)
- Increased Auxiliary Power Unit reliability with higher-altitude starting capability
- Additional cargo compartment fire suppression
- A duplicate electrical system

In the case of the twin-engine 767 and the A300/A310, the aircraft in popular use at the time, both Boeing and Airbus had to make modifications and add capability, whereas the later 777 and A330 were designed from the outset to be ETOPS-compliant.

By 1988, the program had achieved such a high level of success that the FAA and other regulators eased the ETOPS requirement to 180 minutes from an airport, provided the airline was able to achieve a lower shutdown rate of 0.02/1,000 hours.

At the time, Airbus put the reliability of its twin-engine aircraft into perspective in its publication "ETOPS-Twins Through Time", claiming that an ETOPS-equipped Airbus aircraft would have to fly for 8,940 years under the 180-minute diversion criteria before it suffered loss of power from both engines from independent causes. Under the 120-minute criteria Airbus claimed that it would have to fly for 17,880 years-that is, back to the beginning of the Ice Age.

The next stage was termed Accelerated ETOPS and this was introduced in 1995 after 10 years of ETOPS experience. This meant that aircraft did not need to prove reliability by incident-free flying. Now, the reliability of the various systems and processes were taken as proven. This enabled Airbus to deliver its first A330 with P&W engines certified to 90-minutes ETOPS from day one. Boeing followed with the 777 in 1995, which was delivered ETOPS-certified to 180-minutes and its first commercial flight was an ETOPS one across the North Atlantic with United Airlines.

The Boeing-200LR touched down at Heathrow Airport at 1:30pm GMT on 10 November 2005 after establishing a world record for distance traveled – 11,664nm (21,601km) in 22 hours 42 minutes – nonstop by a commercial airplane. The 777-200LR flew from Hong Kong, eastbound over the Pacific Ocean, across North America, and then over the Atlantic Ocean to London. *Boeing*

In Boeing's case, to achieve ETOPS certification for its 777, it drew on ETOPS experience with its 767s and made 160 refinements to the engine systems and 300 design modifications unrelated to the engines, to dramatically improve reliability. The large number of changes to systems other than engines reflects the fact that 95% of all twin-jet diversions are made for factors unrelated to engines or the number of engines. These might be smoke in the cabin, cargo fires, decompression, fuel leaks or a system failure.

From the airlines' and their passengers' perspective, ETOPS has brought about a revolution, with airlines able to better serve markets with more non-stop services using smaller twin-engine jets. A further consequence is that costs and thus airfares are reduced because of more non-stop flights. The evolution of smaller long-range more economical twin engine jets, such as the 767, 777, A300 and A330 has brought about what is referred to as market fragmentation. Fragmentation is where airlines move away from one or both major hubs and serve secondary airports with non-stop flights. In 1984, all trans-Atlantic flights were either 747 (60%) or the three-engine DC-10 and Tristar. Almost all flights were from key hubs such as New York and Chicago on one side, and Paris, London and Frankfurt on the other. ETOPS changed all that, with twin-engine aircraft dominating the North Atlantic today.

The welcoming rays of the second sunrise over Newfoundland. Canada, dance on Boeing's Captain Rodney Skarr, on Boeing's 777-200LR's record breaking flight in November 2005.

Whereas in 1984 there was only one flight a day from Chicago to London with a TWA 747, now there are more than 23 flights from Chicago to 14 different European destinations with 767s and 777s. The 747's share of the trans-Atlantic market is just 4% and more than 140 airports enjoy non-stop trans-Atlantic service.

A similar fragmentation of the market is occurring in the Pacific region, with more than 40 cities having direct services. In fact, the US FAA has granted a further easing of the diversion restrictions by allowing a 15% extension, to 207 minutes, for 777 operators as needed in the North Pacific region.

Boeing's new -300ER and -200LR models of the 777 will demonstrate a capability of 330 minutes ETOPS as authorities move to lift the 207-minute limit. That extraordinary capability has led to the evolution dubbed Long Range Operations (LROPS) or Extended Operations (EO).

Rolls-Royce Trent 500 powers the A340-500 and -600.
Airbus

A340-500 in the colours of Emirates is certified to LROPS standards despite having four engines.
Airbus

777-200 in the colours of Singapore Airlines.
Captain Kevin Tate

CAN WE FLY FOREVER?

The EO initiative will, like ETOPS, advance aviation safety. ETOPS has been so successful in increasing reliability of all aircraft systems, not just engines, that even airlines and aircraft manufacturers that do not require an ETOPS level of redundancy and reliability for their operations or aircraft, are upgrading to those standards to improve reliability.

For example, Airbus has certified its new four-engine A340-500 and A340-600 to 180-minutes ETOPS standard.

The impetus for a new standard occurred in the late 1990s in large part as an outgrowth of the campaign by Boeing and three US airlines to achieve a 15% (27 minute) extension to the maximum 180-minute ETOPS from the US FAA. The 207-minute exemption was critical to the 777 for northern Pacific operations. It would also close a slice of airspace up to 114 miles wide (185km) over the Bering Sea that is more than 180-minutes from a diversion airport when all six alternate airports in Alaska and northern Russia are unavailable owing to poor weather. On average, this happens about twice a year. After some lively debate, the FAA agreed to permit the extension on a case-by-case basis. That decision came into force in March 2000. But the industry wanted something more formal and the US FAA endorsed the request in 2000.

Both the European Joint Aviation Authority (JAA) and the FAA established working groups for LROPS/EO in late 1999 and early 2000. Regulatory panels were set up to determine revised operational standards that could supersede existing ETOPS rules which would apply not only to two-, but three- and four-engine aircraft.

In late 2002, the Aviation Rulemaking Advisory Committee (ARAC), a committee advising the FAA, presented its findings and recommendations for all aircraft, not just twin-engine designs.

Contrary to popular belief, engine reliability as it pertains to twin-engine aircraft, is not the principal problem for the committee's examination of LROPS/EO issues. Instead, the group's focus was more on matters such as fire protection, oxygen systems, pressurization, flight crew rest areas, hydraulic systems and onboard medical capability.

While cargo fire warnings and medical emergencies do not relate to engine reliability, the provision of oxygen does.

Virgin Atlantic A340-600
Airbus

Current regulations require a descent to 10,000ft (3,048mtrs) in the event of depressurization, with some exceptions if supplemental oxygen is carried. This becomes a problem because of the regulatory requirement to consider simultaneous failure of an engine and depressurization. The engines are an integral part of the oxygen system.

To counter this double problem, Airbus is developing an On Board Oxygen Generating System that will enable the A380 and A340 to remain at much higher altitudes, thus saving considerable fuel. Another consideration is the ability of elderly or overweight passengers to tolerate prolonged depressurized flight above 15,000ft (4,572mtrs), a subject of considerable medical debate.

Airbus is also pushing for four-engine aircraft, like the A340 or 747 to cross the Himalayas, shortening flight times between Europe and southern Asia. The dollar value savings for airlines and passengers is considerable for the extended ETOPS routes. For instance, the Boeing 777-200ER with the new polar routes and 207-minute ETOPS, give the aircraft a 301-plus passenger performance compared with 215 to 275 passengers for the conventional routes from New York to Hong Kong. The significant difference in performance between the worst conventional route and the best polar route is an airline accountant's dream — 25,000lb (11,339kg) in extra payload for the 777.

In November of 2003, the FAA issued draft rulings that would extend ETOPS/LROPS/EO operations from an alternative airport from the current 207-minute limit to 240-minutes. The FAA said that Polar Operations to 240-minutes would be approved on a route-by-route basis. In addition, FAA proposed a new diversion limit for specific city-pairs that is beyond 240-minutes but only "available to those operators that have considerable experience with ETOPS, including 240-minutes ETOPS." However airlines would have to demonstrate an IFSD rate of 0.01/1,000hrs compared to the current 0.02/1,000hrs requirement. This is not a problem for aircraft such as the 777 which is demonstrating a rate of 0.005/1,000hrs. Final industry comment is expected in January 2004.

New US ETOPS rules were pending throughout 2006. In January 2006, the FAA sent the new rules for government approvals. The final set of new ETOPS rules from the FAA were ratified in early 2007.

Famous aircraft manufacturer Donald Douglas once said that all advances in aircraft design depend on the engine. And so it was when Boeing set out to design a new aircraft – the 787 – that would be the new benchmark for economy, engines were going to be a critical element. General Electric (GE), the world's dominant engine maker began studies based on its huge GE90 core.
As a baseline, it aimed for:

- A 15% fuel burn reduction compared with its CF6-80 models powering the 767 and A330.
- Pressure ratio close to 50 for the new engine, compared with up to 42 for the GE90 on the 777.
- Bypass ratio 11:1 – well up on the 7.7:1 on current engines.
- Noise goal of Heathrow's stringent QC1 noise limit for departure.

To achieve this, GE employed a host of new and emerging technologies:

- Fan and fan case made from composite materials.
- Integral vane frame and a low single annular TAPS (twin annular pre-swirl) combustor.
- New low pressure turbine design with fewer blades and stators, which offer big maintenance savings downstream. (Stators are fixed vanes that help guide the flow of gas or air through the engine.)

To achieve the ultimate in efficiency, Boeing opted for an all-electric 787 doing away with the need for bleed air off the engine because in current designs, this reduces efficiency.

Aircraft systems, such as cabin pressurization, de-icing of the wings and other environmental control systems, are traditionally powered using hot high-pressure air taken or "bled" from the high-pressure compressor of the engine but this requires a complex array of heavy titanium ducts, pumps and valves, which in turn requires considerable maintenance.

But aside from the efficiency, Boeing and the engine makers were driving the "green" bandwagon. The engine efficiency and additional weight reductions in the airframe meant that the 787 would use 20% less fuel than any other aircraft of its size.

Other green parameters were:

- Target of 26EPNdB (environmentally perceived noise decibels) below the internationally agreed Stage 3 noise levels in force by the early 2000s.
- Nitrous Oxide (NOx) - production 60% below the latest "Chapter 4" levels set by the International Civil Aviation Organization's (ICAO) Committee on Aviation Enviromen tal Protection (CAEP).

Rolls-Royce's first Trent 1000 for Boeing's new 787

CHAPTER 6

Aerodynamics

Aeronautics was neither an industry
nor a science. It was a miracle.
— Igor Sikorsky —

DEVELOPMENT OF THE WING

Aerodynamics is an exacting science, so much so that many aircraft designs have failed dismally - costing billions - because critical drag calculations were out by just a few percent. If drag is too great, the aircraft cannot meet the speed, range and economy guarantees. Even today, with the use of super computers and sophisticated wind tunnels, manufacturers sometimes get it wrong, with disastrous results. In the early years, as you can imagine, aircraft designers employed a great deal more guesswork.

Vickers Vimy, used to pioneer the England-Australia air route took 21 stops and 135 hours and 55 minutes to complete the journey.

The father of aerodynamics was Ludwig Prandtl, born in 1874 in Bavaria. He formulated important aerodynamic concepts, the most notable of which were his boundary layer, thin-airfoil and lifting-line theories. Prandtl evolved his aerodynamic work around 1901 and in 1904 delivered a ground-breaking lecture to the Third International Mathematical Congress in Heidelberg, Germany. Titled "Uber Flussigkeitsbewegung Bei Sehr Kleiner Reibung" (Fluid Flow in Very Little Friction), the paper outlined his boundary layer theory. This theory contributed to an understanding of skin friction drag and how streamlining reduces the drag of aircraft wings and other moving bodies. Prandtl also calculated the effect of induced drag on lift. Induced drag is created by the vortices (turbulent air currents) that trail an aircraft from the tips of its wings where the upper and lower boundary layers meet at the trailing edge of the wing. Vortices affect the pressure distribution over the wing and result in a force in the direction of drag.

Similar findings were also described by English scientist Frederick Lanchester, who published the foundation for Prandtl's theory in his 1907 book, "Aerodynamics". Prandtl also evolved theories relating to thin-aerofoil, subsonic airflow and

Handley Page HP-42. Note cluttered wing and tail design.
British Airways Museum

its effect on the compressibility of air at high speeds, shock and expansion waves in supersonic flow and the effects of turbulence.

Before 1919, many aerodynamicists believed that thinner aerofoils (wings) were the most efficient structures for the speeds of the era, but Prandtl's research showed that the opposite was the case. This led to the construction of much thicker wing sections enabling aircraft designers to insert additional strengthening inside the wing and do away with the external wire bracing that was common at the time. This served to reduce drag and costs.

Prominent in the design of aerofoils was the US National Advisory Committee for Aeronautics (NACA - now NASA) which undertook to test more than 100 designs of aerofoil in wind tunnels, eventually publishing their findings in 1933. Designers were then able to choose the wing section they wanted from the database. To this day, wing designs are sometimes referred to by their NACA number.

One of the other major developments of the era was the "slotted wing" or leading-edge slats which, in simple terms, improved the wing's lifting performance at low speed. Another major development, the trailing edge flap, also emerged during WWI but because of the relatively low speeds of aircraft during that era, its implementation was not extensive. Flaps were found to enable a reduction in the wing area of an aircraft, thereby reducing its weight, and decreasing drag. An excellent example was the DC-1, which after employing flaps, had a 35% improvement in lift coefficient, compared to the competing Boeing 247 without flaps.

Through the 1920s and into the 1930s however, most aircraft designers relied on dramatic increases in engine power for performance improvements and paid scant regard to the innovations promised by aerodynamicists. This fact was highlighted at the time by Professor Melville Jones, of Cambridge University, who claimed that some aircraft of the day had three times more drag than was required for efficient operations. In other words, had they employed the principles of aerodynamics at the time, aircraft could have cruised much faster and lifted more weight. They would then have operated more efficiently with a subsequent reduction in fares.

Germans and Americans however realized the importance of an aerodynamically clean aircraft when they built such models as the DC-1/2/3 and the Heinkel He170 bomber. Streamlining came from a variety of design changes, including retractable undercarriage, engine cowls (covers for streamlining) and flush riveting. In fact, just adding an engine cowl to the Lockheed Vega in the 1930s was found to increase the aircraft's airspeed from 157mph (252km/h) to 177mph (284km/h).

Another area of aerodynamic consideration was burbling or flow separation. Significant drag and thus decreased lift results when the air flowing over the wing breaks away from the surface and becomes turbulent. One trouble spot was the wing/fuselage join. Researchers found that by adding a contoured and streamlined surface to the joint in this area of the aircraft, the problem of turbulence was greatly diminished. In the case of the DC-1 in 1933, a gain of 18mph (29km/h) was achieved after such a modification was added. As the 1930s closed, great effort was put into the dramatic improvements that aerodynamics could generate for both commercial and military aircraft.

COMPRESSIBILITY AND THE JET

During WWII, aircraft only approached the speed of sound during dramatic dive maneuvers, at which time pilots encountered control difficulties produced by the effects of air compressibility. Little research was undertaken into the massive rise in drag associated with the proximity to the speed of sound.

The clean lines of the Douglas DC-3 with DC-1 in the background make a stark contrast to the HP-42 on the previous page.

Albert Betz, however, a Swiss aerodynamicist, postulated correctly that swept-back wings would significantly delay the onset of compressibility associated with supersonic airflow. Again, the Germans led the way and by the end of the war, many designs featured swept-back wings. Certainly swept wings were the answer to high subsonic flight speeds but they also brought some problems.

It was found that as the angle of sweep-back increased, the lift was decreased and the effectiveness of the flaps was reduced. Greater take-off and landing speeds and a higher angle of attack were found to be required, the more pronounced the sweep-back. These problems were illustrated in the battle between the 707 and the DC-8. Boeing went for a 35-degree sweep-back for higher speed, while Douglas concerned about a high landing speed, opted for a 30-degree sweep-back and a taller undercarriage.

When Douglas stretched its DC-8 to carry another 80 passengers, Boeing was unable to economically follow suit because the increased length of the fuselage would have restricted the take-off angle and thus reduced the payload. The only solution for Boeing was to lengthen the 707's undercarriage which would involve a massive and costly redesign effort. Instead, Boeing turned to the 747 to compete with the Super DC-8s.

Actually, Douglas did not have it all their way. In fact, the contrary was the case. According to Harold Adams, one of Douglas' chief designers, the DC-8 wing was a disaster. His explanation, in his book "The Inside Story: The Rise and Fall of Douglas Aircraft", is another lesson in the great hurdles faced by designers as they try to gain a competitive edge. One of the Douglas designers, "Kleck" Kleckler, proposed an aerofoil (wing) designed by the NCAC some years before, which promised lower drag at higher Mach numbers than conventional

The 707 borrowed heavily from Boeing's B-47 jet bomber design, which itself drew from German wing work.

Douglas' competitor to the 707, the DC-8 was a gamble and despite its "clean" appearance suffered from serious drag problems, which were costly to remedy and affected sales.

aerofoils. A conventional wing at 30-degree sweep-back was limited to Mach 0.80, while 35-degree of sweep would give a speed of Mach 0.85.

The wing proposed by Kleck promised to give a speed of Mach 0.85 but with only 30-degree sweep by implementing a "supercritical blunt nose". But the design had never been proven in flight and if the wind tunnel calculations were wrong, which they often were in the 1950s, the problems would be massive because the drag would be higher than that of a conventional wing. Douglas went to the airlines with speed guarantees of Mach 0.84. The unfortunate outcome was that the wing design fell short on guarantees and the aircraft was thus limited to a speed of Mach 0.79 to achieve the range.

A 5% reduction in range meant airlines such as Northwest, switched to the 707, and Pan Am, which had originally purchased more DC-8s than 707s, dumped the DC-8. Complicating matters for Douglas, 147 aircraft had been produced before the vital modifications to the wing were implemented, although a re-fit package had become available to modify existing models.

As aircraft designers attempted to serve small communities in the late 1950s with typically shorter runways, greater use of high lift devices such as leading-edge slats had become necessary. The innovative 727 featured a substantial leading-edge slat and triple-slotted flaps. On landing, the wing appeared to almost come apart, as the flaps fully extended. But the 727-100, which could carry more than 100 passengers, was able to get into and out of airports that the 707 could not have attempted. So technology in the form of advanced wing design was able to bring jet economics and thus lower fares to smaller communities.

While modifications were made to the DC-8 wing the ultimate answer was a lengthening of the wing and redesign and streamlining of the engine pods. These changes, plus two fuselage stretches resulted in the extremely successful Super 60 series. Shown here are the DC-8 Super 62 (above), which was 6.6ft (2m) longer and the Super 61 (below), which was 37ft (11.27m) longer with the wing from the standard DC-8. The third Super 63 variant combined the wing changes of the 62 with the longer fuselage stretch of the 61.

COMPUTERS COME TO THE RESCUE - SOMETIMES

There is no question that advances in computers have revolutionized wing design in the past 40 years. Calculations that once took more than three weeks to complete and check can be performed in seconds today on a home PC.

An excellent example of the improvements in wing design through the use of computer modeling is offered by Airbus Industrie, which claims that its A330 wing is 40% more efficient than the wing of the A300B4 from which the A330 is derived.

But things can go seriously wrong. The performance problems of the MD-11 are an excellent example of the high-stake risks of commercial aviation, even with the help of computers. In December 1986, McDonnell Douglas launched the MD-11, a slightly stretched version of its DC-10, but with an extra 1,242 miles (2,000km) in range. Both General Electric (GE) and Pratt & Whitney (P&W) supplied bigger versions of existing engines to power the jet. This was not a new aircraft/engine combination with a whole lot of imponderables but a derivative of an existing design, yet when the flight-test program began in earnest in early 1990 - nine months late - officials were horrified to see that the specific fuel consumption (sfc) exceeded guarantees by up to 6%.

Boeing's 727 innovative wing featured a substantial leading-edge slat and triple-slotted flaps. On landing, the wing almost appeared to come apart, as the flaps fully extended.
Boeing Historical Archives

A number of key customers with the biggest commitments, Delta, American and Singapore Airlines, had all ordered the MD-11 to perform specific range/payload missions. The 6% hike in fuel consumption meant the aircraft could not fly from Singapore to Paris nor could it fly from Dallas or Atlanta to Tokyo or Seoul with the required payload. At the time, McDonnell Douglas had orders and options for 344 MD-11s from 30 customers but the performance problems resulted in the company building only 200 with virtually no airlines converting options.

The MD-11 saga is also a reminder of how easily an aircraft program can get out of control when management is distracted. Early in 1989, as the first MD-11 was nearing completion, Chairman John McDonnell implemented a Total Quality Management System (TQMS) in which all 5,000 Douglas managers had to reapply for their jobs, with evaluations by both peer and subordinate staff members. The effect on management was shattering and they were hopelessly distracted as a critical program was being prepared for roll-out. McDonnell was frustrated that Douglas could not deliver aircraft on time and, while everyone agreed that change was needed, the effect of such measures was savage.

Cleanlines of the 727
Boeing Historical Archives

Strangely, the problems at Douglas related to production, not engineering and sales, but all were affected. Singapore Airlines cancelled its commitment in 1991 and ordered the A340 instead, due to the prolonged doubts about the performance of the MD-11.

Ironically, the story has a strangely aerodynamic twist. Faced with massive penalties, not to mention cancellations, Douglas engineers set about attempting to claw back the MD-11's performance. In what became an exceptional effort, Douglas and the engine makers were able to increase the range of the aircraft by nearly 10%, actually exceeding the original specifications. Engineers examined the aircraft in minute detail and were able to streamline the airframe, improving range by nearly 5%. One interesting example was the aircraft's cockpit windshield contour, which was modified slightly giving a valuable 0.3% improvement in range.

The MD-11 in flight test. At the time, the aircraft had orders and options for 344 but serious performance problems resulted in cancellations and only 200 were built.

One innovation for aircraft designed in the 1960s has been the introduction of the winglet, which reduces the negative effect of drag on the wing. Designers found that drag could be reduced by increasing the span of the wing or the vertical height of the wing. These devices reduce wing-tip drag by dissipating the vortices created by the merging airflow over the upper and lower surfaces of the wing. Winglets have been a great success on aircraft such as the 747 and 737, which were originally designed in the 1960s.

More modern aircraft, such as the 777 designed with the help of super computers, do not typically need winglets. On the 737NG, the latest version of the 737, for instance, winglets have resulted in a 4% reduction in fuel burn and a 149 mile (241km) increase in range.

WEIGHT ON THE DESIGNERS AND THE WINGS

There are numerous loads that designers must take into consideration when designing a wing. Such variables as maneuver loads, gust loads and landing loads must all be taken into account. There are hundreds of different load conditions to be analyzed.

One critical consideration in the structural design of the wing is the placement of the engines, which provide what is called bending relief for the wing. The biggest challenge for wing designers is to minimize the upward force on the wing created as the wing flies and generates lift for the aircraft. If engines are mounted on the wing, as in the A340 or 747, the weight of those engines helps counteract the upward force on the wing. If the designers place the engines at the rear of the aircraft, the wing must be strengthened.

Likewise, placement of fuel will greatly affect wing up-bending and careful placement of fuel can provide bending relief during flight. On a typical long-range aircraft, the centre fuel tanks are used first and the wing-tip tanks last to provide as much downward weight as possible during the flight. On a 747, the fuel load is typically about 380,600lb (173,000kg) with 246,400lb (112,000kg) in the wing tanks.

Note the wing bending up on Boeing's 747-400 at lift-off. *Boeing*

CHAPTER 7

Testing, Testing

In flying I have learned that carelessness and overconfidence are usually far more dangerous than deliberately accepted risks.

— Wilbur Wright —

Steam roller "torture test" of the DC-1 wing
Boeing Historical Archives

ON A WING AND A PRAYER

The challenge of verifying the performance capabilities, reliability and structural integrity of aircraft in the early years of aviation required incredible innovation, which often bordered on the bizarre.

When French flying boat builder Rohrbach-Romar wanted to test the wing integrity of its flying boat in 1928, there were no such things as test rigs and computer modeling, so a couple of hundred productions workers standing on the wings had to suffice. There were no calculations on whether this was adequate but it seemed good enough and the aircraft's wings were never torn off.

In 1932, while aircraft design had advanced considerably, structural testing had not, so Douglas was forced to use a steamroller driven across the wing to test the structural integrity of its DC-1 wing.

The strength of a structure such as a wing was one thing but metal fatigue of the aircraft's fuselage was another. This, however, did not become an issue until pressurization and jet technology were combined in the Comet in the early 1950s.

Pressurization was introduced into aircraft in 1940 with the Boeing 307 but its altitude was limited by its piston engines. Big military jets were flying in the late 1940s but their hours were low, whereas the Comet 1s were flying every day for more than 10 hours. Structural testing for metal fatigue was virtually unknown until the Comet disasters, which struck in 1954.

In the exhaustive tests that followed, a Comet 1 airframe was immersed in a giant tank and water was pumped in and out to simulate the pressure change with altitude. Eventually a tiny crack developed near the corner of a window frame and the airframe failed.

A310 structural wing testing *Airbus*

777 structural wing testing. Note wing (circled) bent up 24 ft above it's normal position. *Boeing*

TESTING A SCIENCE

Today, the testing of a new aircraft is an extremely sophisticated and expensive process. Not only do manufacturers need to conduct structural tests on the aircraft but comprehensive system tests also are required.

When Boeing built the 777 in the early 1990s, the company spent $370 million on an Integrated Aircraft Systems Laboratory (IASL). The role of the IASL was to test the entire aircraft's systems, which were laid out in 46 labs. The plan was that each of the systems would be tested separately and then integrated into the "complete aircraft".

This level of testing was needed to meet the rigorous demands of United Airlines which wanted the 777 to be certified to 180-minutes ETOPS.

The IASL lab tested nearly 60 major aircraft systems and more than 23,000 parts on the 777. Structures such as the undercarriage, which weighed 10,400 lb (4,545kg), were tested 40,000 times. The level of detail in the testing process was intense and went as far as having wiring bundles the same length as in the aircraft. In one of the labs, more than 40 problems were uncovered and solved that would only have been identified once the aircraft took to the air. When all the separate systems testing was done, the System Integration Lab (SIL) was "test flown" twice a day and at its peak was clocking up 400 "very short" flights a day during the latter half of 1992.

777 brakes glow red during the dramatic brake test. *Boeing*

Firemen attend to the 777 tyres that have deflated by design after overheating in the torturous brake test. *Boeing*

TESTING TO DESTRUCTION

While the various labs were systems-testing the 777, Boeing was torturing the aircraft itself in what became the most comprehensive structural test in Boeing's history.

Boeing, like all manufacturers, had to prove that the structure could carry the maximum design load in the most extreme conditions over the life of the aircraft. For that purpose, it tested individual components and then took the second fully-built 777 and tested it to destruction.

The rig to do this looked more like a medieval torture chamber. In the rig, there were 96 hydraulic actuators to apply stress to the aircraft and there were no less than 4,300 gauges to measure the effect of the stress. There were 500 miles (805km) of cables connecting the gauges to the data acquisition system, where about 1,500 channels of data were monitored in real time by structural engineers who validated predictions and identified structural weak points.

The 777 was severely punished in 120,000 simulated flights - or double the aircraft's expected life cycle. Each flight lasted about four minutes and the fuselage was pumped with air to a pressure of 8.6lb per square inch. The flight consisted of taxiing, climbing, cabin pressurization and depressurization, descending and landing. To add realism, Boeing mixed up the flights, from completely smooth and level ones to those that would have the toughest travelers reaching for sick bags.

The 777 came through the tests with flying colors and Boeing was able to report that the aircraft had 40% fewer microscopic cracks than its 767, at that time, considered a benchmark in the industry. The most visual part of the test was the wing testing. After simulating two lifetimes of testing, the engineers bent the wing upwards 24ft (7.31m), which was equal to a load of half-a-million pounds before it fractured.

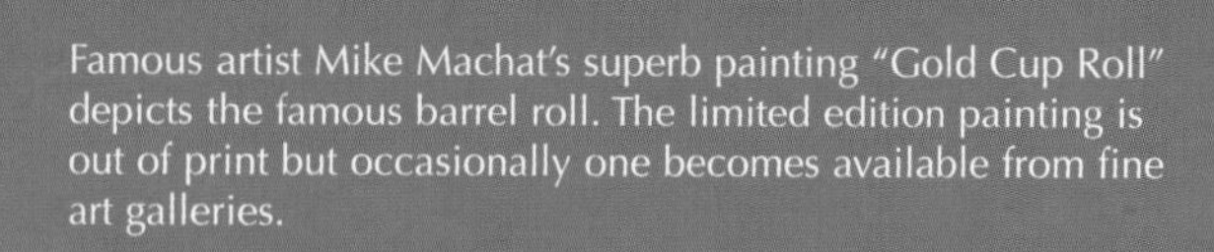

Famous artist Mike Machat's superb painting "Gold Cup Roll" depicts the famous barrel roll. The limited edition painting is out of print but occasionally one becomes available from fine art galleries.

FLIGHT TEST

Having come through the structural and systems tests with flying colors, the next hurdle was the flight test program and, ironically, some of the most rigorous tests actually occur on the ground.

No story about testing an aircraft is complete without reference to the most famous test flight of all. This was performed by Boeing's legendary test pilot "Tex" Johnston. During the 1955 IATA annual general meeting in Seattle, Boeing hosted delegates at the Gold Cup power boat races on Lake Washington. The highlight was to be a fly-by of the Dash-80 707 prototype but to impress the chiefs of the world's airlines, Tex put the Dash-80 into a 1G roll which, while not overstressing the airframe, gave Boeing's then-President Bill Allen severe heart palpitations.

Not content with one roll, and in case any of the airline executives thought they were seeing things, Tex brought the Dash-80 around again and repeated the maneuver. The next day Tex quipped to Allen when asked about the barrel roll, "I was just selling airplanes."

Testing the 777 was a mammoth effort and involved nine aircraft over a two-year time span as three different engines were certified. Typically, an aircraft takes 12 months to be certified. The initial 777/Pratt & Whitney combination was certified in 10 months and one week.

Photo taken from the Dash 80 (707 prototype) when it was performing the famous barrel roll in 1955.
Boeing Historical Archives

One of the most impressive tests is the velocity minimum unstuck (Vmu) which determines the aircraft's minimum lift-off speed. The test pilot rotates the aircraft under the normal take-off speed with a result that the aircraft "sits" on its tail before lifting off.

Another torturous test is the rejected take-off test. The 777 is loaded to its maximum take-off weight and brought up to take-off speed. Without using reverse thrust, the pilot must bring it to a stop using brakes only. To add to the difficulty of the test, the 777's carbon brakes have to be degraded to the point of replacement and after stopping, the brakes become so hot they glow red, but no action can be taken for five minutes, representing the time taken to get emergency vehicles to the aircraft.

Other tests involve Boeing searching the world for the worst weather conditions. In Sweden and Alaska, it found freezing temperatures for the cold soak tests, while New Mexico and Alice Springs in Australia provided searing temperatures. For crosswind landings it was off to Alaska again. When the final tests were complete, the fleet of nine 777s had completed 6,700 hours and more than 4,800 flight cycles.

Commercial aircraft today must demonstrate compliance with 4,800 regulations. Each aircraft is fitted with 22 racks of computers that weigh 34,000lbs (15,422kg) and gather more than 50,000 different measurements.

A340-600 Cold soak tests.
Airbus

An A380 equipped with a special tail bumper performs low-speed takeoff tests, known as VMU (Vitesse Minimum Unstick) tests, at Istres Air Base in Southern France. VMU testing demonstrates the minimum speeds at which an aircraft can takeoff with its rear fuselage touching the runway. *Airbus*

Airbus A380 performs the water ingestion test. The tests which took place at Istres Air Base in Southern France involved accelerating and decelerating the A380 to 70 knots through a water trough. *Airbus*

CHAPTER 8

Technology Delivers Safer Skies

It is far better to be unsure of where you are and know it than to be certain of where you are not

— Famous navigators' saying —

THE AIRCRAFT COCKPIT

Aside from the growth in the size of engines that power today's aircraft, there is possibly no better visual example of the development of aircraft technology than the aircraft cockpit.

The first aircraft were rudimentary affairs with just a handful of instruments to provide the pilot with very basic information.
However, on the Boeing 314 Clipper the cockpit was complex and required a crew of five. Even on the first 707s in 1959, you would find a four-crew cockpit, with two pilots, an engineer and a navigator. There was an observation dome for star sightings to assist with navigation.

The navigator's position disappeared in the early 1960s, together with the engineer's role. While the three-engine 727 required an engineer, the twin-engine DC-9 was certified by the Federal Aviation Administration (FAA) with just two pilots. However, this caused conflict with flight-crew unions because the DC-9 was considered a high capacity aircraft and crew responsibility was not well-defined.

In 1981 the FAA deemed that a flight engineer was no longer required on high capacity twin-aisle aircraft and that a two-pilot crew was safe, due to the advances in cockpit automation and the dramatic improvement in engine reliability.

That dramatic increase in automation was underscored when Boeing moved from an analog cockpit with gauges and dials on the 747-300 to a computerized cockpit for the 747-400 with eight-inch Cathode Ray Tube (CRT) displays. The result was the elimination of 600 dials and gauges.

An illustration of the improvement in cockpit technology was the significant reduction in crew procedures from the 747-300 to the 747-400.

Normal procedures:	107 to 34
Engine fire:	15 to 4
Fuel jettison:	11 to 7
Decompression:	20 to 3
Cargo fire:	16 to 2

The simplification doesn't end there. Fourteen computers dedicated to flight control on the 747-300 were replaced by just three on the -400 model, while three Multipurpose Control Display Units (MCDU) replaced 11 radio/navigation/flight-data/load panels and 24 dedicated test switches.

With the MD-95 variant of the DC-9/MD-90 series, later renamed the 717 after the merger with Boeing, a glass cockpit was introduced.

The MD-90 cockpit. Ultimate development of the DC-9 "classic" cockpit with electronic flight instrument system (EFIS) and full-flight management system (FMS).

DC-1 cockpit.

Lockheed Constellation cockpit. *Qantas*

THE ULTIMATE COCKPIT

The safety record of more recent aircraft, such as those within the A330/A340 and 777 generation, is perfect. Other designs with the latest flight decks, such as the 737NG and 717, also have a flawless safety record. While these models are in the early years of the life-span, it is clear that their safety will outclass those of previous designs.

Boeing 777-200ER cockpit
Captain Kevin Tate

What is termed "the glass cockpit" was introduced by Airbus with the A320 family in the mid-1980s, followed by the 747 and MD-11 in the early 1990s. The glass cockpit is also found on all new regional jet aircraft from manufacturers such as Bombardier and Embraer.

Today's cockpits are uncomplicated in layout with a high level of automation. For instance, where starting the engines once involved dozens of procedures, the A330 or 777 can be started with the push of a couple of buttons and a flick of a switch.

Airbus A330 cockpit. *Airbus*

THE COMMON COCKPIT

Commonality is a feature of Airbus' aircraft, made possible by the introduction of fly-by-wire, which is based on the concept of electrically powered rather than hydraulic flight controls. With this feature, Airbus is able to make the handling characteristics of each aircraft almost identical.

A320 Cockpit which started the "star-wars-styled" glass revolution. *Airbus*

Airbus developed full design and operational commonality with the introduction of digital fly-by-wire technology on civil aircraft in the late 1980s. As a result, ten aircraft models, ranging from the 100-seat A318 through to the world's biggest passenger aircraft, the 555-seat A380, feature near-identical flight decks and handling characteristics.

The benefits of commonality for operators include a shorter training time required for pilots and engineers to transition from one type to another. It also leads to savings through streamlined maintenance procedures and reduced spare parts holdings. Within the A320 single-aisle family for example, 95% of the parts are common to all aircraft. Boeing quotes similar figures for its 737 family, and its 757 and 767 models have common cockpits.

Boeing 767 simulator, showing instructors panel to the left *Ansett*

Simplification of cockpits reduces the risk of pilot error, thus enhancing safety. Commonality also gives those responsible for scheduling increased flexibility, as aircraft of different sizes, such as the A320 and the 350-seat A330/A340, can be operated by the same crews. It also enables pilots to fly a wide range of routes - from short-haul to ultra-long-haul.

CROSS CREW QUALIFICATION

Cross Crew Qualification (or CCQ) is a concept developed by Airbus, which enables pilots to change from one Airbus fly-by-wire family type to another, via "differences training" instead of full transition training. The CCQ concept was approved by the FAA in 1991.

For example, pilots changing from an A320 family aircraft to the bigger A330 or A340 need only eight working days for CCQ instead of the 25 working days it takes for full type-rating training courses. Pilots changing from the A330 to the A340 require only three days, and to move from the A340 to the A330 only one day.

These time-savings lead to lower training costs for airlines and an improvement in crew productivity. The annual savings in payroll costs attributable to reduced transition time can be $300,000 annually for each new Airbus aircraft added to the fleet, Airbus claims.

Airbus A320 family. Composite photo of the Air France A318, A319, A320 and A321
Airbus

MIXED FLEET FLYING

Building on Airbus' operational commonality, numerous airlines have implemented Mixed Fleet Flying (MFF). With MFF, a pilot's qualifications can be current on more than one fly-by-wire aircraft type simultaneously. This means that they can regularly switch from short- and medium-haul operations on the A320 family to very long-haul flights on the A340, for example.

Boeing 777 cockpit *Boeing*

The benefits from the MFF concept enable long-haul pilots to switch to short-haul giving them more take-off and landing opportunities and thereby eliminating the need for currency training. Short-haul crews switching to long-haul are subject to less fatigue from intense traffic high-cycle operations. Airbus claims that there is a significant improvement in productivity due to an increase in revenue hours flown by pilots for the same duty hours because there is an associated reduction in standby and dead time. Further, overall CCQ and MFF improvements can lead to a saving of up to $1 million annually for each new Airbus aircraft added to the fleet.

However, Boeing's counter-argument is that 91% of the world's commercial aircraft are flown from standard flight decks using the familiar wheel-and-column controller instead of the side-stick found in all Airbus types since the A320. It further claims that the FAA initial training requirements are from five to eight days shorter for Boeing flight decks than non-standard (Airbus) flight decks.

Some Boeing aircraft do have common cockpits, such as the 757 and 767, while the 737NG has a similar layout to the 777 and 747-400 series. The challenge for manufacturers is to keep commonality but at the same time introduce new technology to the cockpit.

SIT AND FEEL

Ever since people could fly, the importance of training was realized but in the early days, this was more likely to be just sitting in the cockpit and feeling the aircraft controls. Sometimes this was taken further by setting the aircraft's position into the wind to get a better feel of the control surfaces.

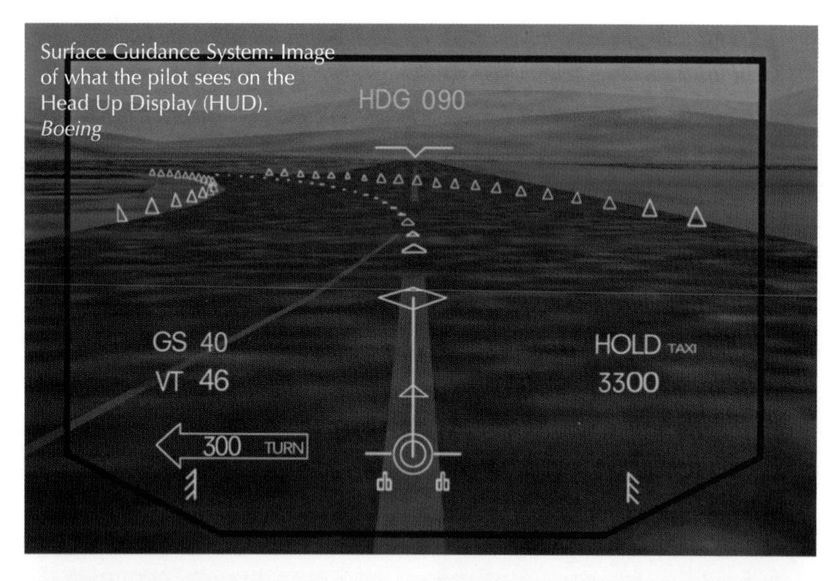

Surface Guidance System: Image of what the pilot sees on the Head Up Display (HUD).
Boeing

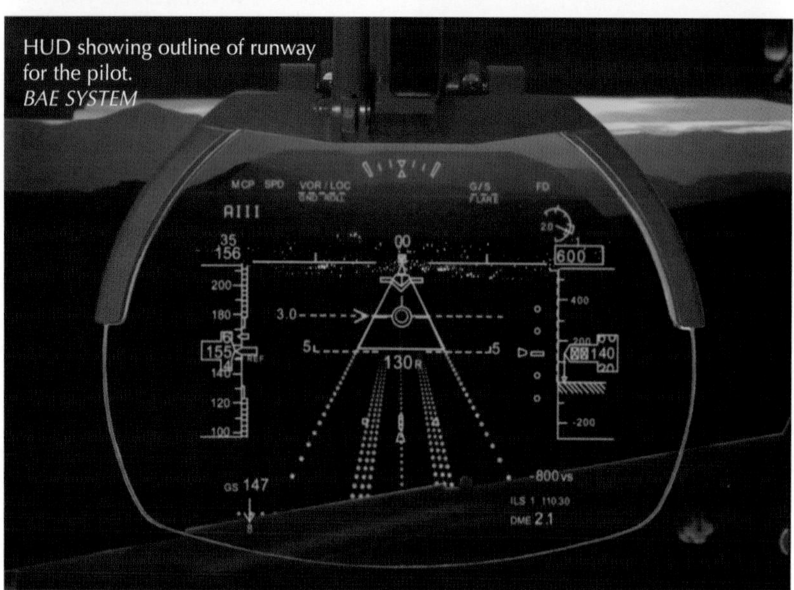
HUD showing outline of runway for the pilot.
BAE SYSTEM

By the end of WWII, commercial aircraft simulators had developed into excellent systems and procedural trainers. A limitation to the early simulators was a lack of analytical information on the performance of the aircraft and engines from manufacturers but the dawn of the Jet Age addressed this shortcoming. At the same time, the second generation of digital computers started to emerge, which met the need for more speed and cost requirements of flight simulation.

The X-15 research aircraft, which flew in 1959, marked an important milestone in the development of flight simulators. It was the first aircraft where the simulator was used to practice flight maneuvers before they were tried in the aircraft.

In 1958, BOAC ordered a Comet IV pitch-motion simulator from Redifon. The simulator was capable of producing motions with two to three degrees-of-freedom. By the time the 747 arrived, simulators had six degrees of motion.

Today, all aircraft manufacturers build cockpit simulators and fly them hundreds of times before the new aircraft leaves the ground. This enables designers to rearrange cockpit instruments and improve information displays, as well as providing detailed analysis of the aircraft's handling qualities. Millions of dollars are saved by avoiding design errors that might normally only become apparent after the aircraft has been built.

Alteon Simulator Bay-Seattle
Boeing

Lightning bolt from a thunderstorm cell around 25 nm away on route to Darwin over the Northern Territory, Australia makes a spectacular image. The photo was taken from an EMB-120ER Brasilia. *Misha Popov*

The Boeing 787 Dreamliner flight deck. All display formats in the 787 flight deck are the same as in the 777, including the flight management computer pages, the overhead (systems controls) layout and the autoflight mode control panel layout. Key features of the 787 Dreamliner flight deck include larger displays, dual head-up displays and dual electronic flight bags. *Boeing*

SEEING IS BELIEVING

Ever since simulators have been around, systems for producing the extra-cockpit visuals have been proposed. However, the realism of today's simulators is only a recent development.

In the 1950s, a black-and-white TV camera racing across a model of landscape around an airport was state-of-the-art. The first color system was produced by Redifon in 1962, while the first computer image simulation was developed by General Electric for the space program. Progress in this area has been dramatic, closely linked to the remarkable developments in digital computer hardware.

The work for the space program was a major factor in the widespread adoption of flight simulators for training. When the Apollo Moon Landing Program was in full swing, the critical role of training in the simulator was widely publicized. One of the key elements, aside from training the moon-landing astronauts, was to train the NASA controllers in dealing with sets of emergencies.

For civilian applications, the simulator not only offered a serious training alternative, it also saved money by not requiring the use of a line aircraft. The simulator also eliminated the need to put aircraft though dangerous scenarios, such as engine-failure training, which cost many lives and aircraft in the 1950s and 1960s. During this period, for some aircraft types, more lives were lost in training accidents than in actual service.

What evolved through the 1970s was what is termed Zero Flight Time Training (ZFTT). This meant that all conversion training was to be carried out either in flight simulators or during routine revenue flight operations. The cost savings were staggering.

SIMULATORS TODAY

Simulators have become a technological entity in their own right capable of duplicating any situation. They afford an incredible degree of realism, right down to the bump on the "runway" from the expansion gap in the concrete. In fact, you can get quite ill from the motion of the simulator.

But the realism had a downside a number of years ago, when airline instructors thought it humorous to program in a series of emergencies that, in real life, were beyond the bounds of mathematical possibility. Pilots quickly learnt to hate the simulator and it declined in usefulness as a training tool until the folly of the instructor's actions was recognized.

Today's airline simulator is an exact replica of the aircraft cockpit and there are simulators for all commercial aircraft models. The only difference is that there is an instructor's station in the cockpit.

The benefits of flight simulators to aviation, the flying public as well as the community, are many.

The simulators are so "real" that authorities accept that they are far better training for pilots than the real thing.

Some benefits are:

- Training for situations such as tire blow-outs, engine failures and stalls, which are not possible in real aircraft because of the danger of losing or damaging the aircraft.
- Ability to train for rejected take-offs which would otherwise put unacceptable wear on brakes and even risk the aircraft.
- Ability to simulate any environmental conditions, such as cross-wind landings, whereas airlines may wait weeks to get the right conditions in real life.
- Pilots can practice landings at any airport on the airline's network before taking the actual flight. This is important for big airlines because many pilots may not visit an airport for some months. British Airways once required pilots to perform a landing at the challenging Hong Kong Kai Tak airport in the simulator before flying the London-Hong Kong flight.
- No disturbance to airport communities from touch-and-go landings.
- Enormous fuel savings.

Simulators have played a critical role in reducing the very high loss rate of aircraft. In the early 1960s, Boeing put the figure at 60 hull loss crashes per one million departures. By the end of the 1960s, the loss had dropped to about five per one million departures and today is about one per one million departures.

RELENTLESS SEARCH

Ever since aircraft took to the skies the industry has been striving to develop new tools to improve air safety, aside from improving the reliability of the basic aircraft systems and engines.

The 777 series has a perfect safety record.

The US provides the best snapshot of the progress of air safety:

1931: Fatal accident rate 0.2938 per million aircraft miles.
1941: Fatal accident rate 0.0335 with the advent of aircraft such as the DC-3.
1951: Fatal accident rate 0.0171.
1961: Fatal accident rate 0.0052 due to weather radar, landing aids and better navigation.
2001: Fatal accident rate is 0.0003.

Today, commercial aircraft such as the 777 and A330/A340 are delivered with a full suite of the latest technology such as Traffic Collision Avoidance System (TCAS) and Enhanced Ground Proximity Warning System (EGPWS), although some levels of sophistication are left to airline choice.

Pilot error at 70%, still remains the major cause of air crashes but it also reflects the number of airlines still operating jets and piston-engine aircraft dating back to DC-3s. Disturbingly, many of the ageing aircraft are operated into hazardous mountainous airfields that lack even basic navigation aids.

Boeing's Statistical Summary of Commercial Jet Airplane Accidents shows that typical first-generation jets such as the DC-8 had an accident rate of 5.87 crashes per one million departures (PMD). Early jumbo jets such as the 747 brought that rate down to 1.97. However, the latest generation aircraft such as the A330/340 and Boeing 777, 737NG and 717 have a perfect record so far.

The accident record of the 717, which started life in 1965 as the DC-9, gives an insight into the dramatic effect that technology has had on aircraft safety.

1965: The initial DC-9 had a crash record of 1.26 PMD.
1980: An updated model, the DC-9 Super 80, later renamed the MD-80, has a crash rate of 0.43 PMD.
1996: The ultimate version, the MD-95, renamed 717 after the merger with Boeing, has a perfect record.

A similar comparison can be made with the A300 which first flew in 1972 and the A330 which started service in 1993. Airbus' A300 had a crash rate of 1.59PMD, while the A330, with its high technology "glass cockpit", has a flawless safety record so far.

WHERE AM I?

"It is better to be unsure of where you are and know it, than be certain of where you are not."

The often-quoted navigator's saying haunts the airline industry as it grapples with the fact that lack of situational awareness is a primary cause of aircraft accidents. According to Britain's Civil Aviation Administration (CAA) Global Accident Review 1980-1996, a lack of positional awareness is the biggest Primary Causal Factor accounting for 24.4% of fatal crashes between 1991 and 2000 and 41.4% when all Causal Factors are taken into consideration.

According to the CAA database, pilot error is responsible for 68.80% of fatal crashes, while systems failure is 3%, maintenance error 1.5%, engine failure 1.3% and aircraft design shortcomings 1%. Accident figures can be misleading because they include statistics from Third World countries. A classic example is Africa where, between 1992 and 2001, an average hull-loss rate of 10.2 PMD was recorded, compared with the average world hull-loss rate of just 1.18 PMD.

The African continent's lack of navigation aids was highlighted by the fact that while Africa accounted for just 3% of world air traffic, it was host to 31% of approach/landing and 50% of Controlled Flight Into Terrain (CFIT) accidents. Worldwide in 2002, CFIT accidents accounted for 18 of the 40 commercial aviation hull losses, including four jets. Lack of situational awareness was a major factor in all of the 219 CFIT accidents that occurred between 1980 and 1996, according to the CAA database.

The greatest single advance in the fight against CFIT accidents is the Ground Proximity Warning System (GPWS). Its installation since 1974 has played a significant part in the dramatic reduction in CFIT accidents from 1.20 per 10 million departures to just 0.35.

However, like all systems it has its shortcomings. It does not function when the aircraft is in landing configuration and provides limited warning of rapidly rising terrain. But there are better models on the way and by July 2005 every passenger aircraft carrying more than 10 passengers must have an Enhanced GPWS/Terrain Avoidance Warning System, according to the International Civil Aviation Organization (ICAO).

Unlike the earlier model, EGPWS has a terrain database, which is used to warn pilots of approaching high ground. This is displayed in the cockpit in shades of green, orange and red.
The pilot receives the information overlaid on this primary flight control display. However, Honeywell Chief Engineer-Flight Safety Systems Donald Bateman and father of GPWS and EGPWS cautions that, owing to the challenges involved in assembling the terrain database from a variety of sources, the system is not yet perfect.

He explains that "we have better databases for Mars and Venus than we have for Earth and we are finding various anomalies around the world, with runway locations sometimes a quarter-mile off, and this is critical in mountainous terrain." But despite the problems, the EGPWS system has an outstanding safety record as can be seen from the table below:

Controlled Flight Into Terrain
Large Commercial Jet Aircraft
Crashes per one million departures

No GPWS	1.20
GPWS-World	0.35
GPWS-North America	0.03
GPWS-Asia	1.10
GPWS-Africa	2.10
EGPWS-Optimum	0.007
EGPWS-Basic	0.01

IF YOU SEE ANOTHER PLANE GET OUT OF THE WAY

In 1919, the US Army Air Service, the forerunner of the USAF, listed 27 flying regulations. Number 12 stated, "If you see another machine near you, get out of its way". There was no argument with that logic but today with the speed of aircraft, if you see a jet, it's almost too late to take evasive action.

Modern cockpit technology in the form of TCAS keeps these jets apart. A Northwest 747-400 and a Singapore Airlines 777-200 cross paths.
Captain Kevin Tate

Today's Traffic Collision Avoidance System (TCAS) is designed to take the guesswork out of avoiding other aircraft, in the event of a breakdown in Air Traffic Control instruction. TCAS was introduced in the US in 1989 and shortly thereafter around the world.

The concept for TCAS dates back to the early 1950s, when air traffic started to grow considerably. It gained further momentum after the 1956 mid-air crash involving a United DC-7 and a TWA Constellation over the Grand Canyon. As a result, the Beacon Collision Avoidance System evolved using aircraft transponders. This was later developed into TCAS in the 1980s.

The FAA launched the TCAS program in 1981 after a series of mid-air collisions due to failure of Air Traffic Control (ATC). TCAS identifies the location and tracks the progress of aircraft equipped with beacon transponders. Basically beacon transponders work by broadcasting an electronic signal to ATC radar giving details of the aircraft. There are various ways that the TCAS information is displayed and on the latest aircraft, other aircraft in the vicinity are shown on the pilot's primary flight display as an open diamond. If an impending conflict is detected, the open diamond turns solid yellow with an aural alert to the crew "Traffic Traffic". If no corrective action is taken the diamond turns red, with verbal instructions to the pilot to climb or descend. The opposite instruction is given to the pilot of the other aircraft. TCAS has saved thousands of lives over the past 15 years.

FUTURE COCKPIT

A host of new technologies are now available that will take air safety to new levels. Most are either certified or are close to certification but none are yet mandated for use on aircraft.

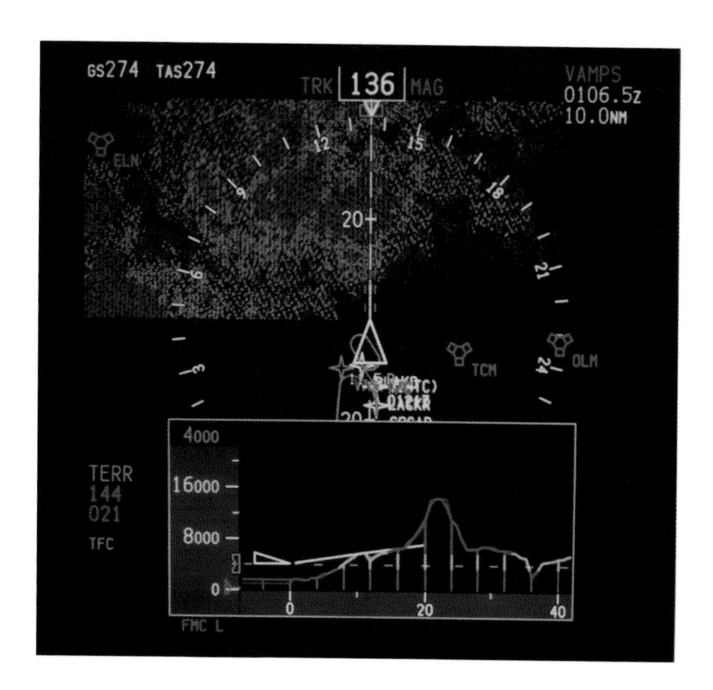

Vertical Situation Display showing terrain ahead of aircraft at the top and the vertical display at the bottom showing the aircraft's flight profile and mountain ahead. *Honeywell*

Key technologies are:

Quiet Climb System (QCS):
This system automatically reduces engine thrust over noise-sensitive areas, reducing community noise and pilot workload during take-off. The system takes the pressure off pilots and airlines when they must adhere to strict noise abatement procedures. Airlines face stiff fines of up to $500,000 for breaching noise restrictions. QCS only requires a software upgrade to the aircraft's Flight Management System and/or auto-throttle system.

Vertical Situation Display (VSD):
This system works with an aircraft's EGPWS terrain database and displays a side view of the aircraft's flight path and, more importantly, its relationship to the terrain below. It significantly enhances safety by showing the aircraft's current and predicted flight path relative to the ground. VSD reduces crew workload and enhances situational awareness by eliminating the need for the flight crew to mentally integrate airspeed, vertical speed, altitude and waypoint information - data available from various instrumentation - into a mental image of the aircraft's vertical position and path trend. Additionally, it helps the pilot determine a stable and appropriate glide path during approach and landing. By indicating terrain well in advance, VSD works with the EGPWS to reduce the likelihood of CFIT accidents.

Navigation Performance Scales:
This allows the aircraft to navigate through a much narrower airspace envelope - possibly between mountains – with greater accuracy. It can help reduce flight delays due to bad weather and also increase airspace capacity. Inertial reference systems, GPS, and ground-based navigation aids are analyzed by the Flight Management Computer (FMC) to track the aircraft's position. The information is displayed in an identical manner to the Instrument Landing Systems (ILS). The system is now available through a software upgrade.

Integrated Approach Navigation:
This is an enhancement to an approach capability, making the pilot interface and procedures very similar to existing Instrument Landing System (ILS) approaches. By allowing a common operational approach procedure, this feature minimizes pilot workload and training, reducing 18 separate approach procedures to one.

GPS Landing System:
This is a highly accurate and reliable satellite-based landing system that will open additional airports and runways to regular service during poor weather conditions, such as heavy fog. This system combines ground-based components with a multi-mode receiver onboard the aircraft. The beauty of this system is that every airport within a 45.9 mile (74km) radius of the ground station has auto-land capability on every runway.

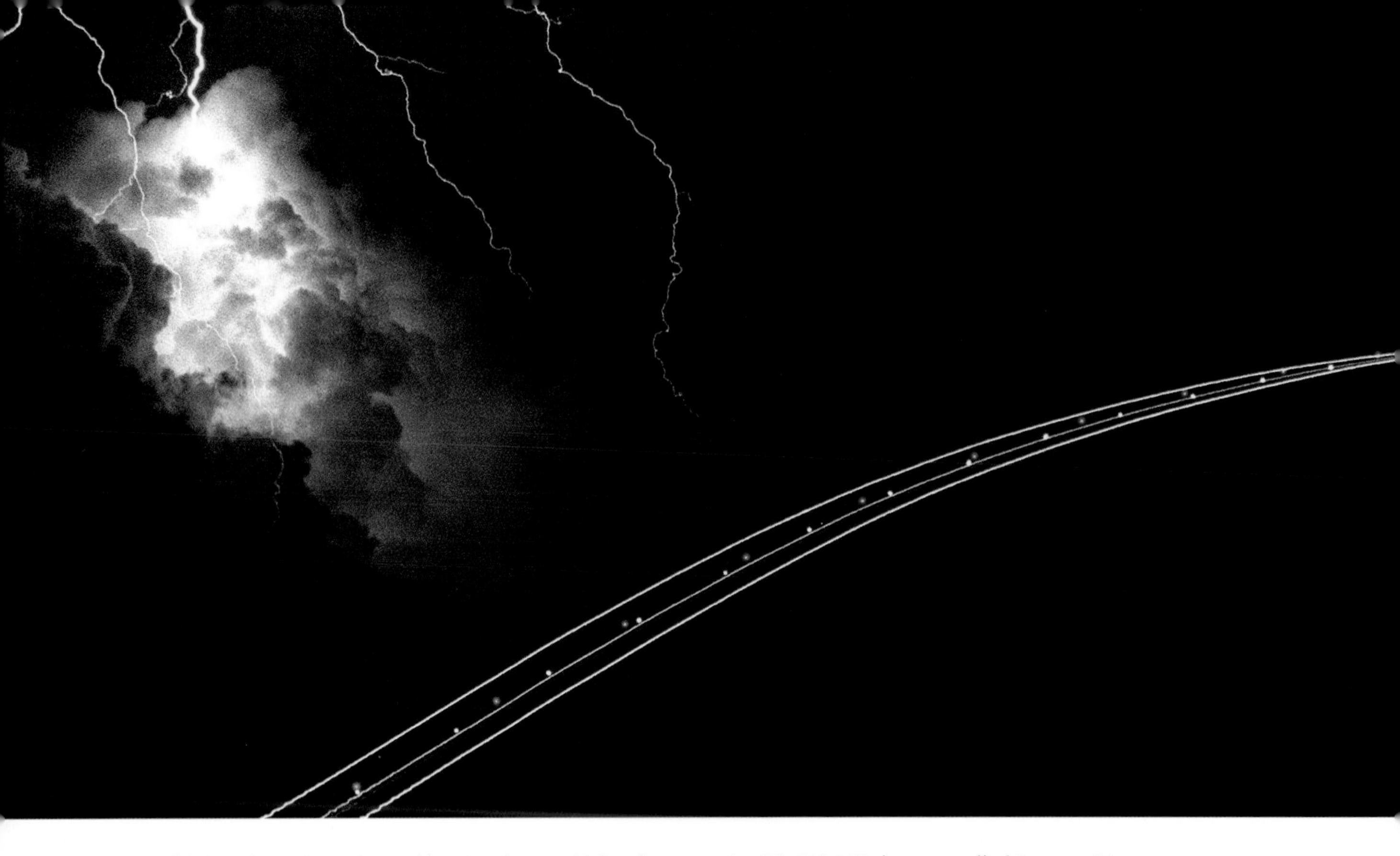

Very sophisticated weather radar enables aircraft to avoid thunderstorms. An ATA 727-200 departure off of Runway 15 at Boston's Logan Airport turns away from lightning in a thunderstorm cell in this 20 second time exposure.
Mark Garfinkel

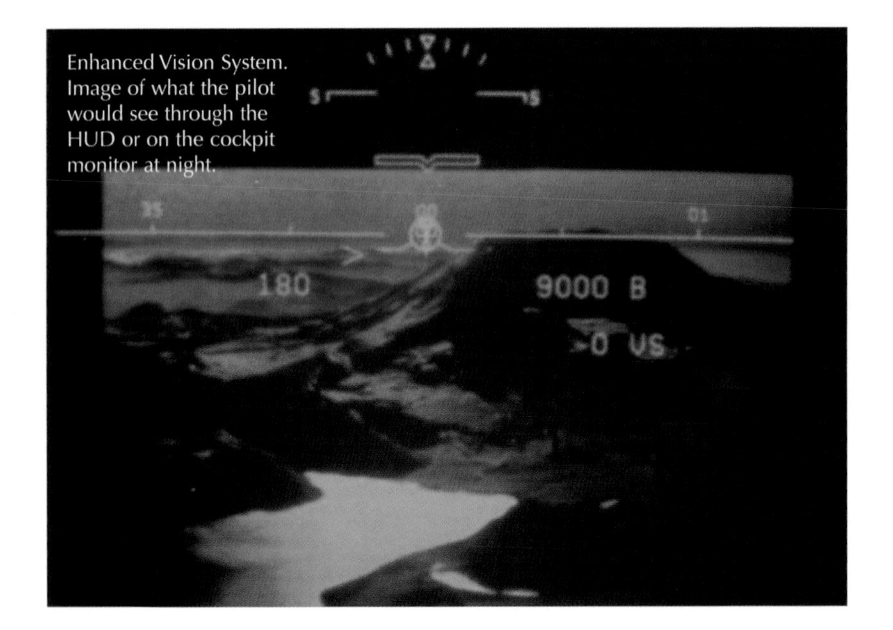

Enhanced Vision System. Image of what the pilot would see through the HUD or on the cockpit monitor at night.

Head-Up Display:
This system, called a HUD, is already in service with some airlines such as Southwest, American Airlines, Alaskan and Qantas. The HUD is a Perspex display in front of the pilots' line of sight that displays all the critical flight information, an outline of the runway and an aiming point. This enables the pilot to land in reduced visibility conditions and see all the critical data relating to the aircraft without looking down into the cockpit.

Surface Guidance System:
This is an emerging technology that improves taxi safety and airport efficiency during poor visibility and darkness by reducing the risk of runway and taxiway related incidents. The taxiway is illuminated on the pilot's HUD and an ATC instruction such as not to enter a runway shown as a row of cones across the taxiway. This technology would give the pilot a clear pathway to the runway and is predicted to save many lives.

Enhanced Vision System:
This system, in wide military use and with police agencies uses thermal (infra-red) imaging which allows pilots to see objects at night and in inclement weather, enhancing safety and reducing delays. The system almost turns night into day and is extremely sensitive. Pilots are easily able to make out vehicles on runways or taxiways. The display is shown on a monitor in the cockpit and also can be overlaid on the HUD.

CHAPTER 9

The Shackles are Broken

I'm flying high and couldn't be more confident about the future.
— Freddy Laker, Laker Airways, 3 days before the collapse of Laker Airways, 3 February 1982 —

TRENDSETTING TRIPPE

Pan Am DC-6 which introduced tourist class across the North Atlantic.

It seems incredible today as airlines fight for passengers with sumptuous meals and the latest hi-tech entertainment that there was a time when strict regulations existed as to how many vegetables an economy class passenger could be served.

As we have noted Pan Am founder Juan Trippe pioneered tourist class service in 1952 across the North Atlantic after a four years battle with the International Air Transport Association (IATA). This class added one extra seat across the width of the aircraft but maintained the 40 inch (101cm) seat pitch and for the first time, passengers had to buy drinks. However, the fare was $270 compared to the normal $395.

Economy class was introduced in 1958 and IATA had very strict rules on what could be offered. Seat pitch was reduced to a maximum of 34 inch (86cm) and IATA detailed to the airlines what the passengers would be served, including a limit on the number of dishes, the number of vegetables to accompany the main dish and absurdly the precise definition of what made up a sandwich. But fares went down by a further 20% and Pan Am charged $463 for the New York-London return trip, a significant reduction on the $567 tourist fare and the $783 first class fare.

Pan Am's founder Juan Trippe poses with Douglas Aircraft Company founder Donald Douglas. *McDonnell Douglas*

LIVE-WIRE LAKER

While Trippe had battled IATA, his battles were nothing to those of Sir Freddie Laker some 20 years later. After the war, Laker who served as an engineer and pilot formed Air Charter Limited, to operate passenger and cargo charters. He also started Aviation Traders Engineering Limited (ATEL) to buy war surplus aircraft and convert them for civilian use.

Sir Freddie Laker hams it up prior to the first trans-Atlantic flight of his Skytrain DC-10.

The story of Laker's "big break" has nothing to do with the thrust of this book but it is a great story. In the 10 October 1977 issue of Time, Sir Freddie revealed that a chance meeting in an English pub in 1948 with an old friend to whom he had once given valuable business advice, was his turning point.

The friend asked what he could do for Laker and "with my usual cheek" Freddie recalls, he told him that he would like £38,000 to buy some aircraft. The man wrote the check on the spot, allowing Laker to purchase twelve converted Halifax bombers from BOAC along with hundreds of tons of spares. A few months later the Berlin Airlift began. Laker sold six aircraft and operated the other six making 4,000 flights in the 15-month emergency.

All his earnings went into buying more aircraft and spares, and by the end of the airlift Laker owned some 100 aircraft and 6,000 engines. He sold some aircraft and smelted down the rest into ingots for saucepan manufacturers. Next his company turned to building aircraft. One of his greatest achievements - and also his greatest failure - was the 28-seat turbine powered Accountant.

ATEL also converted surplus DC-4s into the car-carrying nose-loading Carvair which was a great success.

Sir Freddie Laker accepts another DC-10 from then President of Douglas Aircraft Company John Brizendine. *McDonnell Douglas*

Laker's low fares put massive pressure on older airlines such as TWA. *Art Brett*

Laker sold out but remained in the industry and subsequently became managing director of British United Airlines which was formed in 1960 after the merger of some of his former companies. He built the airline up to be the largest UK independent airline but left in 1965 to form Laker Airways.

Laker Airways was built on the back of the growing inclusive tour market from the UK to the Mediterranean and in 1971, the company purchased DC-10s which set the scene for a collision course with IATA because he wanted to use them to fly across the North Atlantic. The ensuing 2,295-day battle prompted the Duke of Edinburgh to write:

Freddie Laker,
May be at peace with his Maker,
But he is persona non grata,
With IATA.

Laker, from the "wrong side" of the airport, had a massive struggle against the traditional "establishment" airlines. It took him $1.5 million in legal fees and eight hearings to get approval to fly the North Atlantic with both the UK and US Governments intervening to finally support the concept. The approval when it eventually came started a domino effect across the globe.

Laker was charging $236 for the London-New York round trip, compared to $562 for economy and $1,178 for first class on other airlines. Pan Am, British Airways and a host of other airlines dropped fares to try to match Laker and then moved to offer the same level of discounting around the globe.

Some, such as Air India, offered all the frills at the same price in two-page advertisements in the New York Times. Gradually Laker's "Skytrain" service added some frills and evolved into a two-class service and by 1981 was flying nine routes between the UK and the USA. He also ordered 10 Airbus A300s and applied for no less than 666 Intra-European routes.

But conspiracy was in the air and in 1982 Laker was forced into receivership and was peremptorily liquidated. The fact was that Laker was not bankrupt – nothing like it.

Laker Airways filed a $1.5 billion anti-trust action in the US claiming that 12 major international and US airlines and others had conspired to put the airline out of business.

The lawsuit was settled out of court for $56 million of which $48 million was used to pay all the airline's creditors and 14,000 of its ticket holders.

Laker A300
Airbus

Sir Richard Branson with A340-600
Airbus

In true British fashion, as one knight fell another took up the cause to protect the innocent or at least provide them low fares. However, the origins of Sir Richard Branson's Virgin Atlantic were unusual. A London barrister Randolph Fields conceived the idea of an all-business-class airline and brought in former Laker executive David Tait to put the concept together.

They subsequently knocked on Branson's door for equity. The airline was to be called British Atlantic Airways and operate from Heathrow to John F. Kennedy airport in New York.

Again there was a wall of opposition. Plans quickly changed to an economy/first class airline flying between London's Gatwick and Newark in New Jersey. The named changed to Virgin Atlantic Airways and the first flight took off on 22 June 1984.

Sir Richard provides a wonderful insight into the breaks you can get, when he recounts: "On the test flight for CAA approval, one of the engines on the 747 suffered a compressor stall just after take-off and flames shot out of the engine. A journalist took a series of photos and I thought we would be finished with the bad publicity. But the CAA inspector said, change the engine and we will give you your certificate tomorrow and the journalist gave me the roll of film!"

Eighteen months later the airline was granted Gatwick-Miami and Gatwick-Boston followed. Next on Branson's radar was Heathrow Airport and in 1988 he won rights to serve JFK, Los Angeles and Tokyo all from Heathrow.

Virgin's first 747s were based at Gatwick, south of London and frame two British Airways 747s.

Virgin Atlantic A340-600

KLM MD-11
KLM

GLOBALIZATION

According to the paper titled Fettered Flight by Yergin, Vietor and Evans "while globalization is often attacked or lauded as a 'thing' it is in fact a 'process' and is the accelerated integration and interweaving of national economies through trade, investment and capital across historic borders."

The almost irresistible move toward a global open economy is being fuelled by a change of mind around the world away from the concept that governments should control the strategic sectors of a national economy. Airlines of course were very much part of that philosophy. But during the 1970s deregulation in the US, liberalization of the North Atlantic routes and privatization of major airlines such as British Airways brought about a different way of thinking. It was thought at this time that consumers could be better served by competition rather than regulation.

Interestingly the structure of the airline system today has its roots in the Chicago Conference of 1944 where the post-war airline system was laid out. Much of its direction and policy was framed by the war—-projection of power and control of the air. On one hand the US with dominance in the air wanted an open and competitive system for air travel, while Britain and Europe, devastated by war, feared that the US would dominate the peacetime skies as it had done in the war years.

Britain and Europe prevailed and countries had to negotiate air service agreements on a country-by-country basis. The British model did play an important role in protecting fledgling airlines of smaller countries and airlines in many Third World countries.

However, many of the airlines required massive government subsidies to stay in the air and the level of protection given helped to foster gross inefficiency in many cases. Even in the US, regulation reached absurd lengths but some larger states abandoned price controls for all intra-state flights. Texas and California led the way and Southwest, Air California and Pacific Southwest Airlines showed that lower fares would stimulate traffic and increase load factors.

That example led to massive pressure to reform the US airline system and de-regulation followed in 1978. The US case was the seed for change around the globe and governments reassessed their roles in national airlines and regulation of airfares and routes.

The Embraer 50-seat 145 has proved an outstanding success in the regional jet market, particularly in opening up new routes in the de-regulated and open skies environments. *Embraer*

OPEN SKIES

Another major influence on the aviation industry has been Open Skies, which was pioneered by the US in March 1992 when the US Secretary of Transportation called for Open Skies agreements with European countries. Open Skies between two countries allows their airlines unrestricted access to any route linking the two countries.

Many countries have accepted the Open Skies agreements to the extent that by 2000, 50 countries had such agreements in place with the US. The first country to agree to Open Skies was The Netherlands and its national airline KLM.

KLM was an equity partner of Northwest Airlines and was granted "anti-trust immunity" or clearance to collaborate by the Department of Transport, to code-share on key routes. Despite the fact that this would eliminate competition, approval was granted which set the scene for wide-ranging developments of global alliances and equity partnerships.

In the early 1980s, Europe had no less than 104 airlines compared to the 27 major airlines in the US. Many of the European airlines were propped up by governments. Globally, between 1991 and 1997, more than $11 billion was distributed in subsidies for airlines.

In 1995 the EU Commission ruled that the EU central control in Brussels had the power to act on airfares, thereby shifting airline policy from European states to the EU. A series of measures were introduced through the latter part of the 1990s that eventually led to a single aviation market with full traffic rights within the EU by 1997.

Building on the success of its smaller regional jets, Embraer has launched larger 70-95 seat models. *Embraer*

THE EXTRAORDINARY PARADOX OF AVIATION

De-regulation, Open Skies, Globalization and Liberalization have altered aviation forever but they are very much in conflict with government ownership, social agendas and national pride, not to mention the protection of national interests.

An Air Tahiti Nui A340-300
Airbus

These conflicts manifest themselves in a variety of ways across the globe.

- On one hand the US vigorously promotes Open Skies but steadfastly refuses to allow foreign airlines to carry any US domestic passengers. It also does not allow foreign airlines to own more than 25% of a US airline.

- Australia and New Zealand at the other extreme allow a foreign company to setup a domestic airline in their countries. In both cases their respective governments elected to allow foreign companies access to the domestic market with the aim of shaking-up the airline industry and breaking union strangleholds that protected inefficient work practices. Complicating matters for Australia's national flag-carrier Qantas, is that while the Australian Government allows 100% foreign ownership of domestic airlines, it will not allow Qantas to have a foreign shareholding higher than 49%, thus restricting its access to foreign capital.

- Allowing foreign owned airlines to compete on domestic routes can have a negative impact on smaller communities. Typically, a new airline will cherry-pick the best routes, which can benefit the majority of the population, but remote communities can suffer. This is of particular concern in sparsely populated countries such as Canada and Australia.

- Governments continue to be concerned about the vital role that airlines play in linking communities both on the domestic and international fronts. Malaysia is a classic case where the government is keen to ensure that outlying communities get regular service, while at the same time expecting the airline to fly many unprofitable overseas routes to wave the flag and project the country's interests. Indonesia, which is made up of 13,000 islands is in a worse situation with air services critical to outlying areas. However, the government airlines, Merpati and Garuda, while required to serve many unprofitable destinations must face competition from a host of local start ups on their profitable routes.

- Many governments, particularly from developing countries, subsidize national airlines and fares to support national agendas such as bringing in much needed foreign currency and tourists to prop up the economy. Even the smallest subsidies have a significant impact because as the market grows into the lower economic strata it is far more price sensitive and many passengers will change airlines for $50 on a $1,500 airfare.

- When an airline lands in a foreign country it brings with it its higher or lower labor costs and the incumbent airline must compete on that basis.

- Union movements across the globe have totally failed to address the reality of many of these aspects and have made continually unreasonable wage and conditions claims that have led many airlines to succumb to bankruptcy.

While compelling arguments can be made for broadening the accessibility of airlines to all markets regardless of international boundaries, there are equally convincing arguments that such a scenario would be a disaster for many countries and isolate many communities. Fares would decline more rapidly in larger population centers as large airlines fight for traffic. But there is likely to be an increase in airfares in remote communities as these same airlines abandon unprofitable routes to small niche operators, which do not have the benefit of economies of scale and therefore have to bear much higher costs.

Bombardier CRJ900 regional jet in the colors of Lufthansa *Bombardier*

THE RIGHT TO FLY

Securing the right to fly between countries is complex. International aviation agreements are generally negotiated around what are known as the "Nine Freedoms" of air travel. Freedoms one through five have been recognized by nations since the Chicago Convention of 1944. Six through nine have been identified as additional airline freedoms.

Aircraft such as the 777-300ER are breaking down global boundaries.

The nine freedoms are:

1. The negotiated right for airlines to overfly another country's airspace.
2. The right for a commercial aircraft to land and refuel in another country.
3. The right for an airline to deliver passengers from the airline's home country to another country.
4. The right for an airline to transport passengers from a foreign country to the home country.
5. The right to take passengers from an airline's home country, deposit them at the destination, and then pick up and carry passengers on to another international destination.
6. The right to convey passengers or cargo between two foreign countries, provided the aircraft touches down in the airline's home country.
7. The right to carry passengers on flights that originate in a foreign country, bypass the home country, and deposit them at another international destination.
8. The right for an airline to carry traffic from one point in the territory of a country to another point in the same country on a flight which originates in the airline's home country.
9. The right for an airline to originate a flight at and carry traffic from one point in the territory of a country to another point in the same country.

The 8th and 9th freedoms are also known as "cabotage", i.e., the right to pick up and drop off local passengers traveling within the domestic market of a foreign country.

Carriage of passengers involving more than two countries can get more complex and must be founded on one or more of these freedoms, as authorized in a bilateral agreement between two countries. Open Skies agreements, to a large extent, do away with this complexity and allow airlines to achieve the most efficient mix of local and flow passengers across international networks.

CHAPTER 10

Privatization - Blue Skies or Turbulence?

More than any other sphere of activity,
aerospace is a test of strength between states,
in which each participant deploys his technical and political forces.
— French Government report, 1977 —

British Airways has been a shining example of what a privatized airline should be. *British Airways*

BRITISH AIRWAYS LEADS THE WAY

If ever there was a fairytale airline privatization story, it must be that of British Airways. The airline went from being a pariah in the industry to the absolute epitome of what an airline should be. In the 1980s, the government-owned airline was saddled with a strike-prone workforce, bloated management and huge losses and was considered a joke in the industry.

Its airline code BA stood for "Bloody Awful" around the world but a few years after privatization that had become "Bloody Awesome". Before privatization, it was so bad that BA employees would not admit they worked for the airline at parties because conversation would always get around to bad travel experiences. In 1981 and 1982, the airline losses totaled £240 million.

A major part of the problem at British Airways was the perceived high priority of "flying the British flag" that had developed over previous decades with its forebears — BOAC and BEA. Success at BEA and BOAC in the 1950s and 1960s had far less to do with profit and much more to do with a sense of national pride. The old saying that "the sun would never set on the British flag" was well and truly to the fore at BA.

Compounding that mindset was the fact that many of the staff were actually war veterans and the airline was injected with a military mentality. Additionally, the degree of inefficiency was masked by the fact that, due to its protection from real competition, the airline earned money in almost every year from 1972 to 1980. It was virtually impossible to build a case for reform. In fact, by Public Service standards, BA was seen as doing a great job of providing a service and not costing the taxpayer anything.

One of the key underlying problems was that the 1974 amalgamation of BOAC with the domestic and European divisions of BEA had produced not one unified airline but a hybrid racked with management demarcation squabbles.

Efficiency suffered terribly such that in 1976 BA's level of productivity was 42% less efficient than six of its major foreign airline competitors.

Radical surgery was needed in 1981 – not just to get the airline ready for privatization but to save it from bankruptcy. In a measure of British Airways' fall from respectability in the aviation world, the highly respected Financial Times scoffed that the privatization might lure investors because "every market sports a few masochists."

Lord King was appointed chairman in 1981 and described the medicine as "tough, unpalatable and immediate." Staff was cut from 58,000 to 35,000 in nine months, salaries were frozen for a year, 16 routes were cut, stations and engineering bases were closed and many inefficient aircraft were sold off. But the termination packages were so generous that the airline had more volunteers willing to leave than places available for retrenchment. The cost of downsizing was a loss for the year ending 31 March 1982 of £545 million.

Finally, British Airways was privatized on 30 January 1987. More than 1.1 million people applied for shares and, on the first day of trading, the shares jumped 68%. That support was well justified because after privatization the airline's profits more than tripled, while the stock increased seven-fold from the opening price.

Profitable and with total customer focus, British Airways soared with staff productivity up by more than 70% in the ten years from 1981. The airline became the darling of the London stock exchange and the "pin-up" of what privatization could achieve.

Lord King, with Sir Colin Marshal as Chief Executive, also set about turning the airline into a market leader for innovation, with the famous World's Favorite Airline marketing concept. The airline constantly revamped its in-flight offerings adding beds to first class and, later, to business class across its long-haul fleet, constant upgrades to the meal service and a premium economy cabin.

The BA example was not lost on airlines around the world and government decision-makers took particular note. In 1980 virtually all major airlines outside of the US were in government hands. In the early 1990s, privatization had gathered steam with either full or partial sales underway of many more airlines around the globe.

But the mix of partial, full and non-privatized airlines has challenged the quest for a "level playing field". In fact, in 2003 the so-called level playing field looks more like a relief map of the Himalayas.

Airlines that are fully privatized in many cases are pressured to follow government doctrine or social agendas but still be responsible to shareholders and the stock exchange at the same time. They do this while having to compete in many cases with airlines that are government-owned, where the monetary flow is guaranteed and where the agenda has nothing to do with profits.

British Airways A320 being de-iced at Copenhagen Airport.
Egon Johansen

PRIVATIZATION STUMBLES ON CONFLICTING AGENDAS

The privatization trend was to have devastating results in other parts of the world, where a number of airlines had ended back in government hands after brushing with bankruptcy. Malaysia Airlines was privatized in the 1990s but the Asian Currency Crisis of 1997, which severely affected the exchange rate between the USD and the Malaysian Ringgitt, and also the financial position of the airline's principal shareholder, forced the government to bail the airline out.

After 12 years of privatization, Air New Zealand ended up in government control after the purchase of Australia's largely domestic airline Ansett, threatened to decimate the airline. The far-reaching effects of this are still being played out as this book goes to press.

There is no doubt that this saga, which also involved Singapore Airlines is the best example of how privatization and liberalization of markets can founder. The fiasco has become a textbook lesson to the industry on what can go wrong and is essential reading if one is to fully appreciate the complexity of the industry.

On the surface at least, the three-way tie-up between Singapore Airlines (SIA), Ansett Holdings (Ansett) and Air New Zealand (Air NZ) offered all the ingredients for a first-rate equity alliance. SIA is one of the most profitable and best-capitalized airlines in the world. Ansett enjoyed a reputation as a top-tier domestic carrier winning Air Transport World's "Airline of The Year" in 1987 and had begun to develop an international network. Air NZ was a small but high-quality international player, noted for its engineering excellence, wanting to expand. However, the mixture proved to be a recipe for disaster.

As the drive for globalization has grown, airlines have been embracing cross-border equity alliances in an attempt to achieve competitive global and fiscal strength, while still wanting to preserve their national identities. What evolved in Singapore, Australia and New Zealand is not just an inter-regional problem but a lesson for all airlines and governments and it should be seen in that light.

The story of the SIA-Air NZ-Ansett tie-up is also the story of the collision between

Malaysia Airlines was privatized in 1995 before the crippling Asian currency crisis forced the government to buy back its stake.

liberalization on the one hand and the national need for a healthy and vibrant airline system on the other. Governments must reexamine closely the fundamental role of their national airlines and measure that role against the realities of the world market. Airline analysts say that governments need to revisit the issue of cross-border ownership and reassess the nature of government and foreign control of airlines. Deregulation and liberalization cannot work when governments continue to apply regulatory conditions and then play politics.

Air NZ Boeing 737-300
Craig Murray

Peter Harbison, founder of the Centre for Asia Pacific Aviation in Sydney, Australia, observed in 2003 that in every industry in which it is applied, deregulation ignites a remorseless trend toward consolidation and reduction of the number of competitors. Airlines are different from other industries, he notes, because of the combined influences of politics, nationalism and a liberal dose of patriotism. Prevented from full equity mergers by national citizenship requirements they must embrace alliances and marketing collaboration with other airlines instead.

The seeds of Ansett's eventual demise were sown in the Australian Government's decision in the early 1990s to merge its two state-owned airlines: wholly domestic Australian and wholly international Qantas. Overnight, Qantas which had not been permitted to fly within Australia gained a fully developed domestic system.

Ansett, on the other hand, was forced to build its own international system from scratch, a handicap it never overcame. The answer to Ansett's plight and Air NZ's desire to expand was an equity alliance, which was consummated in 1996 when Ansett's 50% owner, Thomas Nationwide Transport (TNT), decided to quit the airline business and Air NZ jumped in to take up the TNT holding. Air NZ saw this as an opportunity to finally gain a foothold in the Australian domestic market, which it considered vital to its future growth, after the Australian Government had reneged on an earlier agreement for unrestricted access to the Australian market.

Ansett A320
Airbus

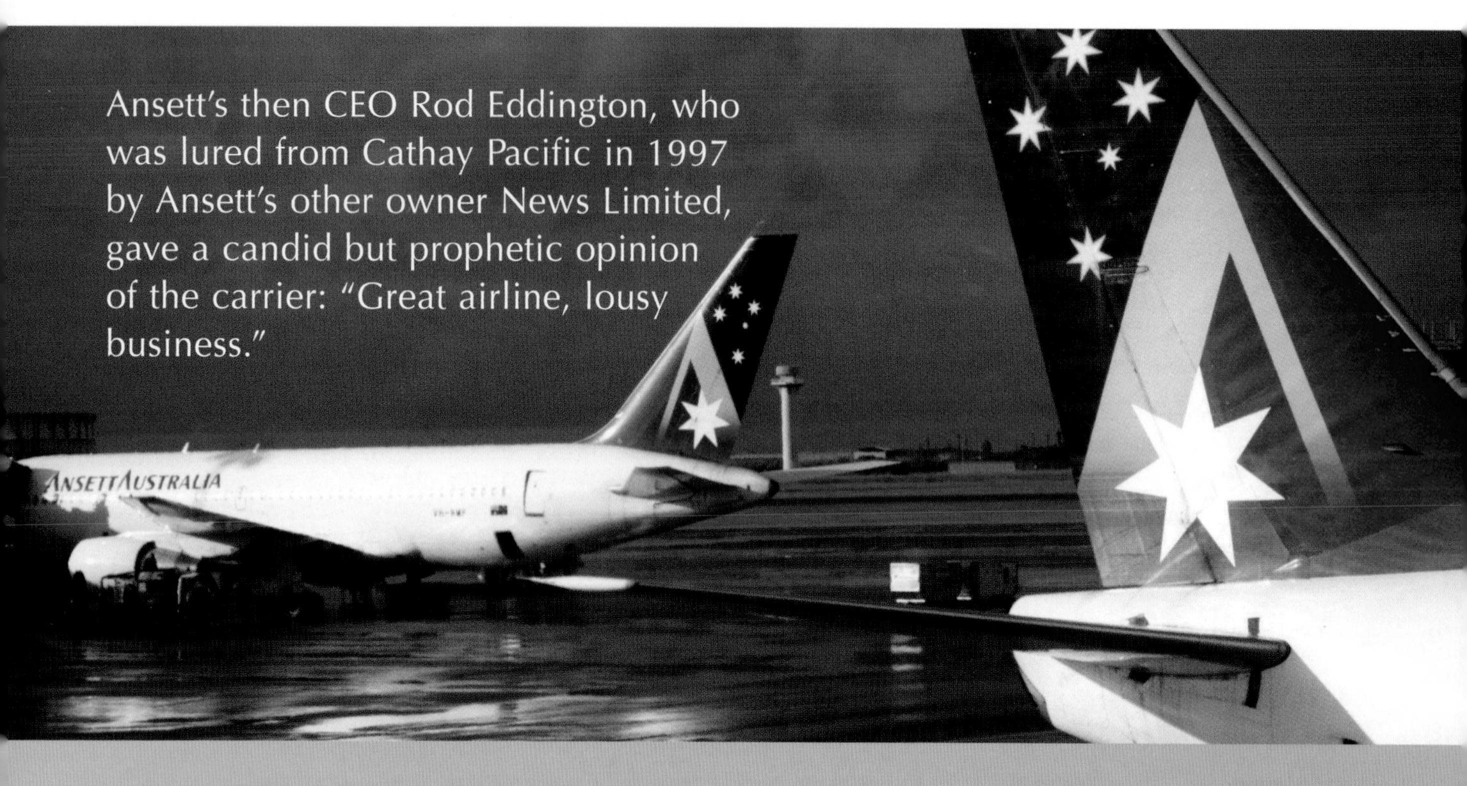

Ansett's then CEO Rod Eddington, who was lured from Cathay Pacific in 1997 by Ansett's other owner News Limited, gave a candid but prophetic opinion of the carrier: "Great airline, lousy business."

Ansett's problems were:

- High labor costs - almost double those of new low-cost entrants
- Inflexible labor unions
- Culture of excess and arrogance developed over two decades
- Thirty-seven subsidiary companies, many unrelated to the airline industry
- Old and mixed fleet with six basic aircraft types
- Lack of desire by principal Air NZ owner Brierley Investments Limited (BIL) to invest in new aircraft
- Declining share of high yield corporate market

Despite these complex problems, within two years Eddington had Ansett in profit and in an alliance with SIA. And against the odds, Ansett was also developing an international presence. Now, however, a new challenge arose.

News Limited was eager to quit the airline business. As a condition of its agreeing to buy out TNT, Air NZ had received a right of first refusal of the News Limited stake. Eddington, however, believed that Air NZ did not have the balance-sheet strength to buy Ansett so he approached SIA, which twice before had tried to secure stakes in Australian airlines -Ansett in 1991 and Qantas in 1994.

SIA sought advice from its bankers on Air NZ's ability to exercise the preemptive right to News Limited's holding. The advice was the same as Eddington's - Air NZ would not attempt to buy the shares, so SIA made its bid for 50% of Ansett in early 1999 and Eddington agreed to stay on for another two years. SIA's then CEO Dr Cheong Choong Kong was delighted. He had Australian government approval and he had access to Ansett International, the airline's international division, which had enormous growth prospects. "It is the jewel in the Ansett crown" he said at the time. Backed by SIA's enormous resources, Ansett International could expand quickly to rival Qantas.

However, Air NZ's owner BIL, now with a 57% shareholding, also wanted out of the airline business and New Zealand's fierce national pride came to the fore. BIL was eager to bring SIA into that airline thereby reducing its own stake at the same time. So Air NZ exercised its pre-emptive rights to the shares and purchased the News Limited stake in Ansett early in 2000.

Australia's Foreign Investment Review Board could have deemed the deal not in the national interest. BIL's position was complex and somewhat removed from both Australia and NZ. BIL, while founded in NZ, was in fact domiciled in Singapore with a significant foreign ownership.

The Australian Government did nothing and the outcome was much as BIL wanted, with Air NZ taking News Limited's share in Ansett and SIA was forced to pursue a back door approach by taking a big slab (24%) of BIL's stake in Air NZ. The outcome in the early part of 2000 satisfied no one:

- SIA ended up in partnership with the company that had blocked its bid for Ansett. SIA, Air NZ paid top dollar for Ansett at a time when the carrier was in need of significant new investment.
- Eddington left because of the new ownership structure.

Summing up the impending disaster Jeanette Ward, airline analyst at global ratings agency Standard & Poors, said that, "It defies belief that they (BIL) didn't know what they were buying." To the surprise of no-one, problems began immediately and:

- Air NZ CEO Jim McCrea was dumped.
- 13 Ansett executives were shown the departure lounge.
- S&P downgraded its rating on Air NZ - the first of four downgrades that year.
- Sir Richard Branson's Virgin Blue and Impulse Airlines started flights with a cost structure half that of Ansett.

As low-cost operators, Virgin Blue and Impulse were tearing at the heart of Ansett in the second half of 2000, the owners of Ansett wrangled over funding for Ansett and ownership of Air NZ with the NZ Government. At the core ofthe problem were the following:

- BIL was apparently unwilling to provide any new funding to Air NZ.
- SIA was not going to sign any blank checks without a bigger stake and control.
- Under NZ law, SIA as a foreign airline was limited to holding no more than 25% of Air NZ.
- New Zealand's Labor government under Prime Minister Helen Clark was not eager to take up the issue of helping BIL.
- Neither the New Zealand nor Australian governments recognized the severity of the growing crisis.
- For most of 2000, Air NZ lacked a CEO with airline knowledge and staff morale was shattered.

The combined effects of these problems, along with the grounding of Ansett's fleet of 767s in December 2000 and again in April 2001 for technical reasons, spelt doom for the airline. Qantas entered the fray in May with a novel idea of taking SIA's stake in Air NZ with Ansett to be sold off to the Singapore-based carrier. Its motivation was to create two equal blocks, Qantas/Air NZ and SIA/Ansett.

Ansett International 747-300. Ansett International was the "jewel in the Ansett crown." *Ansett*

In May 2001, Ansett was hemorrhaging $9 million a week, its yield had dropped 16% and a loss of $200 million for the year to June 30 was tipped. Qantas made its pitch but Air NZ endorsed the SIA proposal of trying once again to increase its shareholding in Air NZ, subject to due diligence on Ansett's health. What unfolded were months of procrastination and backroom dealing as the Australian Government pushed the Qantas line and a divided NZ Government was in a terrible quandary of how to deal with the looming failure of its national airline and its Australian subsidiary.

The Air NZ board was divided and throughout this time the airline's shares were devastated by almost daily changes in direction by analysts' views of government policy direction. Air NZ-Ansett management's entire recovery strategy for Ansett was based on buying out Virgin Blue and eliminating their low-fare competitor, after Qantas had taken over Impulse Airlines a few months earlier.

Sir Richard Branson had indicated a willingness to do a deal in May but had a change of heart as he became aware of positive profit numbers on his fledgling operation. In a typical Branson publicity ploy, he called a press conference in Melbourne Australia, to announce that he had "sold-out" to Air NZ/Ansett and produced a fake check for $125 million, which he proceeded to tear up, saying he "was only joking". But this was no laughing matter for Ansett and Air NZ and the long suffering staff.

Within hours of Branson's rejecting the Air NZ offer, Air NZ had Ansett up for sale to Qantas for just 40 cents. Qantas rejected the 40-cent offer, as did SIA, which by now had a better sense of the grave situation at Ansett.

On September 12, Air NZ put Ansett into administration and two days later, Ansett was grounded. The NZ government, finally aware of the problems faced by Air NZ, agreed to inject $362 million in October and re-nationalize the airline.

In the wash-up, Ansett's unsecured creditors, who were owed over $1 billion received nothing, its staff are still hoping to get 92 cents in the dollar but will probably only receive 55 cents and Air NZ was forced to write off NZ$1.45 billion.

What the collapse of Ansett and near-collapse of Air NZ underscore, is that small countries such as Australia and New Zealand should address in a bipartisan way, precisely what it is they want from their airline industries. Then they should establish a clear strategy to achieve it. This may involve such measures as removing national ownership controls but retaining national board control and reevaluating competition laws relating to mergers.

But that soul-searching came too late for Ansett, which had been flying for 77 years. The airline now stands as a perfect example of everything that can go wrong when governments and industry strive to reposition themselves for a global economy. Another perspective to consider is that airlines in peacetime are more important than a country's air force, and no country would ever get another nation to own and operate its air force.

While the events up to September 2001 were a textbook example of how badly matters can go wrong in the airline industry, the events at Air NZ since that date have been a perfect example of how right they can be when dedicated investors and enlightened management take over. As turnaround stories go, Air NZ is as good as it gets. In October 2001, the company teetered on the edge of collapse as major shareholders Brierley International Ltd and Singapore Airlines baulked at recapitalizing the carrier.

At the end of November, new board member John Palmer took over as chairman and Ralph Norris, one of only two board members to survive the Ansett debacle, was appointed CEO and Managing Director in February 2002.

In just 2½ years under Norris, the airline went from the nightmare of writing off NZ$1.45 billion – the largest write-off in New Zealand corporate history – to posting combined annual profits of NZ$331 million during FY2002 and FY2003.

One of the first things on the agenda under Norris was a staff survey to gauge the airline's heartbeat. The results were disturbing, indicative of a desperate need for some "intensive care" to bring Air NZ back to life. Only 29% of staff bothered to respond, and of those, 90% claimed that management was out of touch, possibly reflecting years of neglect under executives whose entire focus was shareholder return.

Norris's philosophy was different. "It is all about people," he told *Air Transport World*. "The relationship between staff and customers is critical. If you get the staff relationship in good order, then you have a great opportunity to get superior satisfaction. If you get that, then you get superior shareholder returns."

But first he had to stop the flow of red ink which had resulted in net losses of NZ$319 million in FY2002. His measures were harsh but fair and effective. A 30% pay cut was imposed on management and all perks such as chauffeur-driven cars were dropped. Non-management staff was asked to take a pay pause. As most of the layoffs came from management ranks, the staff began to take notice.

A key platform of his rebuilding strategy was to "ensure that people felt their input was valued." And carefully thought-out input was definitely required to rebuild the shattered airline. It was clear the staff was working with tired and outdated equipment and Norris, with a real passion for technology, was determined to give them the right tools. Norris was aggressive in giving staff the right tools. He implemented far-reaching programs and a host of changes, such as:

- The Business Transformation Program, with 40 initiatives to deliver NZ$245 million in savings over three years.

- Radical restructuring of the domestic product with a simplified fare structure was put into place with fares 20% lower on average and some up to 50% below those previously available.

- Elimination of first class on domestic routes lifted cabin seating by 14 to 136 on 737-300s.

- Self-check airport kiosks were installed and within a short time more than 50% of passengers were using this method – the highest take-up rate in the world.

- Overhaul of the airline's trans-Tasman services to Australia.

- Thirty-five A320s were ordered and optioned to replace 767s and 737-300s.

- Boeing 777s and 787s were ordered in 2004.

- A radical overhaul of the airline's long haul product was implemented introducing premium economy.

But not everything went Air NZ's way. Competition regulators rejected a comprehensive alliance with Australia's Qantas to take a 22.5% share in Air NZ worth NZ$550 million.

Air New Zealand's first two Boeing 777-200ERs at the handover ceremony in Seattle in October 2005 *Peter Clark*

The industry believed the alliance was a perfect match but despite government backing, neither the New Zealand nor Australian competition regulators could be swayed. However, both airlines successfully appealed to the Australian Competition Tribunal but a similar appeal to the High Court of New Zealand failed.

The most contentious issues were that the alliance would result in a monopolistic situation with the airlines combining operations across the Tasman and that Qantas would withdraw its services from the Auckland-Los Angeles route.

The historic statistics on fares and wages supported the alliance. According to the Reserve Bank of New Zealand and the NZ Bureau of Statistics, between 1990 and 2000:

- Airfares between Auckland and Sydney declined 47% (based on the lowest fares available from Air NZ),
- CPI had increased 30%, and
- Wages had lifted by 33.7%.

Interestingly, from 1990 to 1997, Qantas had a 19% stake in Air NZ and during that time fares went down by 26%.

In 2006 both airlines revisited an alliance but this time in a more watered down version and as this book closed for press the proposal was before competition regulators.

Air NZ and Qantas correctly argued that by combining flights they could rationalize operations, cut out wasteful duplication of flights, operate larger more efficient aircraft and offer passengers new point-to-point services. These types of services would eliminate the costly and time consuming stops in hubs like Sydney and Auckland. New non-stop services between Wellington and Canberra, the two countries' capitals are also proposed.

Competition watchdogs play a vital role in protecting consumers' interests. However, regulators need to give far more weight to historical downward trends in airline airfares that are virtually identical the world over, regardless of whether one, two or ten airlines compete on a route. The airline industry is unique and governments need to reassess competition policy when making decisions about this industry.

CHAPTER 11

Deregulation - Success or Failure?

Deregulation is the freedom to go bankrupt.
– Alfred Kahn, father of deregulation –

BACKGROUND

The debate on the success or otherwise of deregulation will rage as long as there are economists and universities. Each paper written by an expert builds a compelling case for its success or failure.

Singapore Airlines' Boeing 777-200ER in a pre-dawn arrival at Adelaide, Australia. *Misha Popov*

Many critics suggest that the US airline industry had evolved into a cozy club by the late 1970s. To understand this claim and what led to deregulation, a little history is in order.

In 1938, when the Civil Aeronautics Act was passed in the US, there were 16 major airlines and dozens of smaller ones in the US. Justifiably, government assistance through protection, subsidy, promotion and regulation was thought to be necessary to permit the new industry to develop at the time.

To oversee the industry, Congress created the Civil Aeronautics Authority (CAA) to regulate airlines as if they were public utilities. As such, the CAA was authorized to issue certificates to provide air services between specific points and to approve all fares and schedules. In 1958, Congress ratified the role of the authority but changed its name to the Civil Aeronautics Board or CAB.

By 1978, six of the original 16 major airlines had disappeared through mergers and no new major transcontinental airline had been allowed to start. The remaining 10 accounted for 90% of the air carrier market. However, there was discontent over:

- Alleged price fixing
- Low industry load factors
- Lack of flexibility of the CAB to encourage innovative fares
- Inability to persuade the CAB to liberalize fare regulation

In one classic case that underscored the problem, World Airways, a charter airline, applied in 1967 to fly a scheduled service between New York and Los Angeles at discount prices. The CAB "examined the case" for six years and finally dismissed the application. But even if you were a member of the so-called "cozy club," gaining CAB approval for a new route was difficult.
One of the best examples of this was Continental Airlines. The airline had to wait an incredible eight years just to add San Diego-Denver to its network and only gained that after the US Court of Appeals instructed the CAB to grant the authority. In fact, even if an airline just wanted to deploy its aircraft in a more efficient way, CAB approval was required in order to quit a route.

What did not escape Congress' notice was that small regional airlines such as Pacific Southwest and Southwest, which were not under the control of the CAB because they flew within the state boundaries of California and Texas, were charging much lower fares and were flying full aircraft. Congress also noted the dramatic effect of the relaxing of charter rules, which led to a bonanza for the traveling public with much lower fares.

Deregulation has been blamed for the overcrowding at such airports as Boston International, but the extraordinary growth in air travel has often been seriously underestimated by authorities.
Mark Garfinkel

THE KENNEDY REPORT URGES DEREGULATION

Adding weight were subcommittee hearings chaired by Senator Edward Kennedy. The Kennedy Report stated that deregulation would bring about pricing flexibility, which would in turn stimulate new and innovative alternatives. It would offer passengers a range of prices and services dictated by consumer demand, thereby enhancing airline productivity and efficiency, and increasing the health of the industry.

When Jimmy Carter took office in 1976 the momentum for change had bipartisan support. Carter appointed Cornell economics Professor Alfred Kahn chairman of the CAB. Kahn was a vocal critic of the CAB who claimed it had caused excessive airfares and carrier inefficiency. He assured a wary industry and the public that everyone would benefit. "I am confident that consumers will benefit; that the communities throughout the nation - large and small - which depend on air transportation for their economic wellbeing will benefit, and that the people most closely connected with the airlines - their employees, their stock holders, their creditors will benefit as well."

So in 1978, the Carter Administration eliminated the public-utility model of passenger airline economic regulation and gave airlines the freedom to raise and lower fares and to enter and leave markets at will. The Department of Transport took over the regulatory and competition roles. The results were as dramatic as they were savage.

Many airlines collapsed, fares were slashed in the short term, new airlines entered the market - and soon failed - and tens of thousands of airline employees lost their jobs. Deregulation was a harsh way of dealing with over-regulation.

Lay media lauded deregulation citing the fact that in the 20 years since it was introduced, airfares had dropped 40% in real terms while the number of flights has increased by 50% and passenger numbers soared from 275 million to 581 million. However, those numbers simply reflected the trend that was in place before deregulation and was continuing despite deregulation. In fact, technology was the major factor responsible for driving airfares down.

AIRLINES CRASH AND BURN

One of the strongest critics of deregulation is Paul Dempsey, who wrote "Flying Blind - The Failure of Deregulation." Dr Dempsey claimed that 10 years after deregulation, passengers were paying 2.6% more for fares than would be expected had the decline in airfares continued. He added that routes are more circuitous, service is poorer and there are fewer carriers, not more as promised by Kahn.

A Delta 727 takes-off on Runway 22R at Logan as a US Airways shuttle lands on 27R. *Mark Garfinkel*

The effect of deregulation was indeed savage and the number of major airlines was reduced to just seven - United, American, US Air, Northwest, Delta, Continental and Southwest. Great names such as Pan Am, TWA, Eastern, Braniff, Western, PSA, Republic, Allegheny and Northeast all disappeared. A host of start-ups, such as People Express, went the same way.

Airline profits, which were not solid before deregulation, have been worse since, say some analysts. Many, like Dr Dempsey, argue that airline transport is too critical to the productivity of the economy and the well-being of citizens to abandon it to private concentrations of market power.

Some of the results of deregulation have been:

- The age of the US airline fleet has more than doubled.
- The decline of fares has been no more dramatic than would have been expected when compared to the historical downward trend.
- Fewer of the major airlines have continued operating.
- The development of the hub-and-spoke system has led to increasing distances and time to final destinations.

Some proponents argue that there are actually more major airlines operating, albeit regionals, and combined with the rapid expansion into new markets by many of the established airlines, deregulation has resulted in unprecedented competition. In the US today, 85% of airline passengers have a choice of two or more carriers, compared with only two-thirds in 1978, but one must acknowledge that the natural expansion of air travel may well have brought about the same results.

The same applies to the claims by the Brookings Institute, which in 1999 estimated that the traveling public was saving more than $20 billion a year as a result of deregulation. That figure was arrived at by attributing 55% of the savings to lower fares and 45% to an increase in flight frequency, which, it says, has reduced the number of nights that travelers must spend on the road. While these figures appear convincing, they are actually no different from the numbers that could have been produced at any other time period in airline history as airfares have declined relative to inflation and average weekly earnings every year. Nor are they any different from any other part of the world – much of which has never been de-regulated.

MORE AIRLINES?

But many continue to argue that the benefits of the free-market environment are undisputed. One noted US trade journal declared that US passenger enplanements (passengers boarding) more than doubled in two decades since deregulation to almost 600 million. But they doubled everywhere else as well - without deregulation.

Cargo was also touted as a winner with a trebling of tonnage - but again that occurred around the globe and was more to do with the explosion of the small-package market than with deregulation. Others tout the number of new airlines which has risen from 43 to more than 90 as a positive outcome - but most of those are small operators while many of the major airlines have been forced to merge or have foundered.

A Continental MD-80 taking off from Boston's Logan Airport. *Mark Garfinkel*

The facts are:

- In 1980, there were 10 major trunk airlines, six local service or regional airlines and 44 other airlines – made up of regional, cargo or Hawaiian- or Alaskan-based airlines – for a total of 54 airlines.
- In 2000, there are 15 majors (but this number includes four freight-only airlines), 39 nationals and 46 regionals totaling 100 airlines.
- The doubling of airlines was more a result of a restructure of the industry from "origin-to-destination" flights to "hub-and-spoke" with many new feeder airlines evolving. An example was American Eagle, a division of American Airlines, which made it into the "major league" with revenues of more than $1 billion despite only having small regional jets and turboprop aircraft under 50 seats.

A 757 about to touch down in fog and light rain
Mark Garfinkel

Northwest DC-10-40 poses with a rainbow. Northwest Airlines has survived deregulation by absorbing a host of smaller US carriers such as Republic, itself an amalgamation of more than three airlines.
Mark Garfinkel

LOAD FACTORS AND HUBS

Absurd in some commentaries is the claim that airlines have become more efficient just because of deregulation quoting the fact that airlines are operating at load factors not believed possible 20 years ago. They clearly do not give appropriate weight to the enormous role of computers, yield management software and internet bookings in lifting load factors.

While Kahn says that he is "proud of the role I played in airline deregulation" when he was chairman of the CAB, he qualifies that by saying - "at the same time, it is important to recognize that problems exist, and the task remains one of attempting to grapple with those problems within the philosophical constraints of the free competition deregulation unleashed."

At an Aviation Week and Space Technology conference, Kahn cited what he calls a "disturbing decline" in the average number of airlines per route. He also noted that some airports complained that they are virtually controlled by one airline due to the dominance of the airline at that airport. Of great concern to Kahn and others is that some bigger airlines stymie new competition by predatory practices such as dumping capacity. Kahn cited the example of an incumbent airline increasing the number of discount seats it offered on one route from 1,500 to 5,000 to combat a new "upstart". When the new airline was driven out, the number of seats reverted to 1,500. For Kahn the answer is not re-regulation. He believes that authorities instead should make "every effort to encourage the contestability of markets."

One significant problem for the airline system has been the emergence of what is called hub domination or fortress hubs in the US, as airlines have merged or been taken over. Notable are Atlanta, where Delta has an 80% share, Cincinnati, where it

A Virgin Blue 737-800 at rest. Virgin Blue has been a runaway success in Australia after the collapse of Ansett. *Craig Murray*

controls 94% and Salt Lake City, where it has 90%. Charlotte, N.C. is another where US Airways has 92%, and Pittsburgh, where it has 90%. Continental dominates Houston Intercontinental with 80%, while American dominates St Louis with 70% and Dallas/ Fort Worth with 69%. United dominates Denver with 69%. Consumers, however, are not complaining about low-fare champion Southwest which dominates two Texas airports, Houston Hobby with 81% and Dallas Love Field with virtually 100% of slots.

However, authorities are concerned that the entry of new airlines, believed to be encouraged by deregulation, has all but dried up with only New York-based JetBlue starting up in the past few years. They argue that airports have entered into long-term gate and facility arrangements with major airlines, all but locking out new entrants.

While authorities express concerns, the reality is that investors may have finally realized that the economic return from the world's least profitable industry is all but non-existent. One way around the dilemma of new airlines failing to make inroads into the industry has been achieved by a move in the US in 2003 to increase the level of foreign investment in US airlines from the current 25% to 49%.

East-West Airlines F-28. East-West tried to break the regulation shackles in Australia by flying transcontinental via Ayres Rock but was taken over by Ansett in the mid 1980s.

DEREGULATION - PARADOXES, COMPUTERS AND FREQUENT FLYERS

Commentators suggest deregulation is full of paradoxes. While passengers demand better seats, less crowding and better meals, they will take the lower fare. At the same time, passengers are offered a host of innovative low fares but they find the complexity of fare structures a nuisance.

Early morning arrival of a Malaysia Airlines 777creates some fancy lights. *Misha Popov*

In fact, central to many of the advances attributed to deregulation is the dramatic advance in computers, particularly with respect to reservations systems, which has expanded the industry's core capabilities. Computer reservations technology can be traced back to 1967, when American Airlines conducted an experiment in retail distribution placing terminals in a small number of travel agencies. Progress initially was slow and by 1976 American had installed its SABRE system in less than 150 travel agency offices.

By today's standards, the systems were primitive and they provided flight information and made reservations only for participating airlines, whereas today the systems can host many airlines and link many airlines and related products. The effect of the new level of technology and the subsequent advances in hardware and software has meant that strategic developments in pricing have been possible. Sophisticated yield management systems have led to airlines such as American having up to 540,000 fare levels across its network, while on one day during the peak of fare wars in the US there were 80,000 fare changes each day.

Responding to the competition from upstart airlines, American Airlines introduced loyalty programs in 1981 whereby passengers gained loyalty or frequent flyer points for traveling on American. Within five years, most airlines had adopted such programs. And the passengers were not the only ones getting rewards - travel agents gained higher commission once they steered a sufficient volume of traffic towards the airline.

Frequent flyer programs have evolved over the past 20 years into an extremely important marketing tool that established airlines have used to retain loyalty against low-fare competition. An example of how important the programs are was seen in Australia, after the collapse of Ansett, where 70 billion frequent flyer points were lost. When a rescue of Ansett was mounted, it failed after the proposed new owners abandoned the obligation to honor the points bringing about a huge public backlash.

Boeing 737s being readied for delivery to low-cost operators. *Boeing*

AIRLINES EXIT WHILE SOME PROSPER

Most US airlines were not ready for deregulation and the entry of upstarts, mostly funded by business people who had no idea about the airline industry, was a recipe for disaster. The names of the fallen read like a Who's Who of aviation history - Pan American, National, Braniff, Eastern, TWA, Frontier, Republic, Western, Ozark, Piedmont and Continental. Most of the new start-ups collapsed, the most notable being People Express, but you can also add Muse Air, New York Air, Spirit and Vanguard.

There is no doubt that one of the big winners in deregulation was American Airlines. In the years immediately following deregulation, American quit its older fleet, reduced staff and restructured its route system to build its hub-and-spoke connecting complexes and directed 79% of all traffic to Dallas Fort Worth, where it relocated its headquarters. It introduced aggressive marketing and pricing strategies, including the frequent flyer program, plus Supersaver fares, as well as increasing investment in its SABRE computer reservations system.

American's then-president, Robert Crandall, understood the implications of deregulation better than most and rapidly restructured his airline to meet the greater disciplines required. The results were stunning and between 1976 and 1983, American trebled in market value, while its market share and value have continued to grow. In 1980, American trebled in market value, while its market share continued to grow. In 1980 it carried

Carried Over 1,000 US airlines - large and small - have flown into the sunset since deregulation through merger or bankruptcy.
Mark Garfinkel

25 million passengers, well behind Delta, Eastern and United, all of whom carried about 34 million. Within eight years, American was carrying 72 million passengers, the highest number in the US and Dallas Fort Worth was the second biggest hub in the country.

In summary, regardless of the arguments for and against deregulation the facts are indisputable:

US PASSENGER GROWTH 1950-2005 (Millions)

	Domestic	International	Total	% INCREASE
1950	17,468	1,752	19,220	
1955	38,221	3,488	41,709	117%
1960	56,351	5,906	62,257	49%
1965	92,073	10,847	102,920	65%
1970	153,662	16,260	169,922	65%
1975	188,746	16,316	205,062	21%
1980	272,829	24,074	296,903	44%
1985	357,109	24,913	382,022	28%
1990	423,565	41,995	465,560	22%
1995	499,000	48,773	547,773	17%
2000	610,600	55,550	666,150	22%
2005	670,152	68,273	738,426	10%

Source: ATA

Fact 1: There is no question that airfares have come down since 1979 in the US and air travel has boomed. But this is true also for the rest of the world where most markets have not been deregulated, although many have been liberalized.

Fact 2: The downward trend in airfares since deregulation in the US is no different from that experienced before deregulation, as you will see in Chapter 15.

Fact 3: It can be seen from the figures above that in fact the increase in air travel has slowed since deregulation, leaving aside the dramatic effects of the fuel crisis in the early 1970s.

However the spectacular growth rates of the 1950s and 1960s were all related to the introduction of economy class and the widespread introduction of jets.

Fact 4: In the US since 1979 1,017 airlines - large and small - have being declared bankrupt or been forced to merge.

CHAPTER 12

Aircraft Manufacturing Productivity

All attempts at artificial aviation are not only dangerous to life but doomed to failure from an engineering standpoint.
— editor of 'The Times' of London, 1905 –

DC-8 production line 1958. Because manufacturers are well advanced with mass production before flight tests start, performance shortfalls can be extremely expensive to rectify.

ADVANCES IN AIRCRAFT MANUFACTURING

The cost of buying an aircraft for an airline is huge but the cost of developing an aircraft program for the manufacturers is astronomical. The most significant challenge for aircraft manufacturers in the development of an aircraft program is the length of time before the manufacturer sees any return on its investment. The wait is made worse by the level of inherent risk while manufacturers await reactions to their new programs by the airlines.

Typically it takes four years from the formal launch of a new aircraft program before the first aircraft is delivered to a customer. Then the manufacturer must sell and deliver about 400 of the aircraft before the costs of the development program are repaid and the manufacturer "breaks-even". However, if the slightest imperfections in aircraft performance arise, program costs can sky-rocket and the prospect of "breaking-even" fades, as costly modifications are made to rectify performance shortfalls.

As an illustration of the risks involved, Douglas' DC-8 jet program lost money despite selling 556 aircraft because the aircraft did not meet guarantees. Many variants were built and the company faced expensive production costs because of parts shortages during the Vietnam War and a rapid build-up in production in the mid sixties with the Super 60 variants of the jet. In fact, since the advent of jet aircraft, there have been very few profitable programs. Luckily for Boeing, all their models have exceeded break-even status.

Launch costs of a new program, particularly for manufacturers, can be staggering. In 1952, it cost Boeing $16 million to build the prototype 707, which amounted to about 22% of the company's asset base. That was often described as the $16 million gamble and urban legend had Boeing Chairman Bill Allen "betting the company". While he may not really have bet the company on the 707, he certainly did on the 747, when development costs topped just on 90% of Boeing's asset base.

In an aircraft's sticker price, development and manufacturing costs, according to Boeing's figures, are:

- Materials 30%
- Research development and testing 13%
- Labor 24%
- Program management 16%
- Engines 17%

Unless manufacturers can program a steady increase in production, utter chaos can and has resulted. In 1965 when production of the DC-8 was just a trickle and the DC-9 was just starting up Douglas Aircraft Company had a combined production line.

However within a year DC-9s were selling like hot cakes and required dual production lines with many aircraft being finished off in the company's car parks.

The story was the same for the Super DC-8s.

Boeing's giant 747/767/777 assembly halls at Everett, Washington - world's largest building. *Boeing*

The Increase In Cost Of Aircraft Development Programs Over The Years

Aircraft Model	Manufacturer	Delivery Year	Cost-then year dollars
DC-3	Douglas	1935	$1.5 million
Constellation	Lockheed	1943	$6 million
B707	Boeing	1952	$16 million
B747	Boeing	1969	$750 million
A380	Airbus	2005	$15 billion

Graph 12.1 Source: Company records

As can be seen in the Graph 12.1, the cost of building an aircraft has risen dramatically over the last century. But to put this into perspective, the figures must be considered in the context of aircraft weight and, more importantly, productivity.

For example the productivity of the A380, when measured in available seat miles per hour will be 100 times higher than the DC-3, according to Airbus.

In order to deliver these advances, technological improvements have brought about sustained and sometimes spectacular progress in the industrial capacity to build aircraft.

An interesting insight into the escalating demands of designing an aircraft over the years is provided by the number of man-weeks required to get from the concept stage to the ready-for-production phase.

According to Peter Brook's (B.Sc.) outstanding series, Transport Aircraft Development published by Air Pictorial in 1985, to design the first transport aircraft required about 1,000 man-weeks and the DC-3 in 1933 needed four times that number. For the piston-engine aircraft after 1945, 70,000 man weeks were required. Three times that number (210,000) was required for the first US jet transports in 1955 and one million man weeks for the 747. It is hard to believe therefore that today's programs are able to remain viable. The answer is the huge advances in manufacturing technology.

Boeing executives predict that the company will be able to reduce the time taken to build a 747-type aircraft by two-thirds by 2010. In addition, he believes that Boeing will slash the cost of building such an aircraft by 50% in the same time frame. The reduction in both factors is a product of the rapidly changing influence that technology has had on commercial aviation manufacturing, which is a more recent innovation.

Aircraft engineers see a whole new era in manufacturing technology with advances by a factor of five and in some cases 10. Breakthroughs in information and manufacturing technologies are giving manufacturers the opportunity to turn away from old concepts. Those breakthroughs include advanced design, digital pre-assembly and construction techniques such as three-dimensional solid modeling, high-speed machining, lean assembly and much higher parts commonality.

As an apt example, the design of the 767-300ER raked wing tip has 65% fewer parts than the comparable 747-400 winglet.

This results in:
- 70% less tooling cost
- 44% less fabrication time
- 30% lower cost overall

These new processes are now being transferred to the 787-Boeing's new 250-seater that promises a 20% reduction in operating costs over comparable aircraft. Airbus has a similar story with its new A380 mega jumbo.

Underlying technology making these improvements possible is the use of the state-of-the-art CATIA 3D computer design tool with solid modeling techniques. The CATIA enables engineers around the world to design aircraft without conflicts of systems becoming an issue. If, for instance, a Japanese engineer in his Tokyo office routes an electrical bundle through a space occupied by some toilet plumbing planned by a British designer in Bristol, a conflict alert is flagged and both engineers are advised to sort out the problem before parts are ordered.

An intriguing element of the 3D design is a computerized "virtual" person, which is tasked with performing routine maintenance that will be required on the actual aircraft. If the "CATIA person", which is manipulated graphically on the computer screen, is unable to do so because there is a conflict with another component, or the space is physically inaccessible, the problem is flagged for the designer to make the necessary modification to rectify the situation and design in more accessibility.

LEAN MANUFACTURING

This has become the catchphrase for a revolution in aircraft manufacturing that came to a head when Boeing tried to double the production of its best-selling 737s. Deliveries of parts were late and customers were furious as Boeing had to close the production line down for a month to catch up, costing billions of dollars.

The problem was simple. Boeing was using 1945 methodology to produce aircraft and manufacturing became bogged down in customer options. The classic example was the 737 nose landing gear, which required 464 pages for the optional parts that are attached to each aircraft, and the now famous 109 shades of white paint that Boeing offered. The result was a $4 billion write-off and many disgruntled customers.

In 1998, Boeing set targets of an improvement in quality of 50% a year, a productivity increase of 2% a month and a reduction in lead time for parts of 90%. These goals have largely been achieved. By late 1999, Boeing had reduced inventories by 62% and the amount of space needed by the company's facilities by 50%. For the 737, the time required for production has been reduced from 24 days to just 13, crane movements are down by 40% and the inventory of parts is down 42%.

A good example of lean manufacturing is that of Goodrich, which now delivers to the 737 facility at Renton, 32 fuel probes in three kits instead of the 32 boxes it once did. In each kit, the parts are packed in the order a mechanic will need them. The same supplier also delivers fully built-up landing gears to Boeing instead of the Boeing mechanics putting them together. That has led to a 66% reduction in the number of mechanics required by Boeing for this particular process.

Lean manufacturing has had a major impact on the production of the 777. The latest variant the 777-300ER has taken 40% less time to build than the first -300 aircraft. And that was accomplished despite the fact that 60% of the -300ER used new or revised parts and involved the installation of overhead crew rest areas for the first time.

This success is not only confined to Boeing but is industry wide. At Lockheed Martin, lean manufacturing has reduced production time for the F-16 fighter from 42 months in 1990 to just 22 months, while the cost of building the fighter has dropped by 50% despite declining orders.

As a result of "lean manufacturing" efficiencies instituted over the last ten years, it is expected that the advanced fabrication and assembly processes will:

- Slash tooling time by 90%
- Decrease manufacturing time by 66%
- Reduce manufacturing costs by more than 50%
- Diminish part counts by 50%
- Cut build time from 15 to five months

If there is one aspect that reinforces that the 787 is a game-changer, it can be found in Nagoya, where in the space of a few minutes, you can see both the 787 and 777 wing box constructions.

While it may take just five minutes to get from one facility to the other, it felt like one is going back in time 50 years. The 787 facility could double as the set for the next James Cameron space adventure Avatar while the 777 production hall was a relic from the last century.

Gone was the ear shattering riveting guns, replaced by the "very Japanese" musical tunes to warn that an automatic guided vehicle (AGV) was moving silently across the floor – that incidentally one could happily eat one's next meal off – carrying yet another giant one-piece composite section.

And around the globe there were similar AGVs – all with different tunes or warning chimes – moving in and out of autoclaves and medical-styled clean rooms carrying composite structures which will eventually come together like a finely tuned orchestra.

But therein lays the challenge – for Boeing no longer made the instruments for the orchestra.

All players brought their own and Boeing was left in the role of conductor or designer and integrator – for good reason. This "787 orchestra" pushed technology like no other commercial aircraft since the advent of the 707 and Boeing had turned to what it terms the "best of the best" around the globe for innovation.

Boeing, under former President Alan Mulally, reversed disastrous production problems in 1997 to evolve savvy supply management techniques, including early supplier involvement in aircraft design. It was also about each supplier taking total responsibility to design, build and solve problems – although that problem solving was often done as a collective.

While many saw the 787's "new age" manufacturing processes as a gamble, Boeing's executives did not see it that way at all. Scott Strode, VP of 787 Development and Production, saw it very differently.

"Because of advances in technologies such as composite materials, existing facilities could not accommodate either the kind of work, or the amount of work, that comes with a program like the 787," he said.

All told, there were 3 million square feet of facilities being built to support the 787 production. However, the new age 787 manufacturing also included shifting the responsibility for micro-design and integration to the 1st Tier – major – suppliers, leaving Boeing to undertake the overall design and integration. It also shifted the responsibility for quality control to these suppliers. Balancing that, the 787 suppliers were world's best practice or longtime partners of Boeing – and in many cases both.

Boeing's Scott Strode VP of 787 Development and Production faces the camera with the first center wing box lower panel at Fuji.

Fuji Heavy Industries (FHI) – longtime Boeing partner and supplier – was responsible for the entire center wing-box and also the integration of that center wing-box, whereas Kawasaki Heavy Industries (KHI) was responsible for the landing gear wheel-well. KHI was also involved in the construction of some fuselage barrels, keel beams, pressure bulkheads and the fixed trailing edges. Completing the Japanese involvement, Mitsubishi Heavy Industries (MHI) built the main wing-boxes.

While the plants were expansive, the numbers of workers was small, replaced by robotic automated lay-up machines, non-destructive inspection (NDI) machines, contour tape lay-up (CTL) systems and giant autoclaves.

Half a world away, near Taranto in southern Italy, Alenia Aeronautica had built a massive 753,473sqft facility – enough to house 24 US football fields – at Grottaglie. Alenia built the 787 stabilizer, the fuselage mid-section and the center-aft section.

The passion for the 787 work overflowed. "This program is in our heart," said Giovanni Bertolone, CEO of Alenia. "We are delivering advanced composite technology and this fantastic adventure is a new way of working together." The extent of Alenia's involvement was evidenced by the fact that the company had sole responsibility for design, construction and testing of the 787's horizontal stabilizer.

Alenia's 787 involvement did not end in Italy. They formed a JV with rival Texas based Vought to form Global Aeronautica, located at Charleston South Carolina, with responsibility for fuselage integration. That facility and Vought's own 787 fuselage barrel plant sat side-by-side.

The integration plant took the fuselage sections from Alenia and the numerous parts from Japanese suppliers, which were flown in by specially modified 747-400 Large Cargo Freighters (LCFs), and combined these with sections built by Vought into an almost "complete" fuselage and then flew them to Seattle to be married to the wings and nose sections.

Putting together the 787 jigsaw supply chain was almost as challenging as designing the aircraft itself. According to Strode, Boeing had taken a long hard look at FedEx and Wal-Mart, who made complex logistics look easy.

Aside from the delivery of high visibility fuselage and wing sections there were dozens of suppliers shipping parts across the globe – and with a twist. For instance, instead of SAAB of Sweden delivering cargo doors to Boeing, it delivered the forward cargo doors to Spirit in Wichita, which built the forward fuselage and nose, and the rear cargo doors to Alenia, which built the rear sections.

The world's largest autoclave is used to bake the wings of the 787.

MEGA JUMBO, MEGA PROBLEMS

Singapore Airlines A380 in the final assembly hall at Airbus.

"It is Airbus as a whole which failed [and] the management on several levels who failed. It is only by facing the reality in all honesty that we can overcome the problem." – Christian Streiff, Airbus President and CEO (4 October 2006).

The "problem" was a two year delay in delivery of the world's largest passenger plane – the 555-seat A380 – due to a breakdown in integration of the design effort which resulted in wiring incompatibility. That wiring related mainly to the cabin seats and the complex personal entertainment systems.

Production problems are nothing knew. Douglas was forced into a merger with McDonnell in the mid 1960s because its production lines became snarled, while Boeing had to stop production of its 737 and 747 in 1997 for one month because it had tried to ramp up production too fast.

Airbus told customers in June 2005 that there would be a five month delay in delivery of the A380 due to problems with the electrical harnesses (wiring) and followed that up a year later with an announcement of another seven month delay.

Streiff was then appointed to lead Airbus and conduct a comprehensive review of the A380 program. According to Streiff, the issue of the electrical harnesses is extremely complex with 330 miles of cables, 100,000 wires and 40,300 connectors.

He explained that the "root cause [of the A380 wiring issue] is that there were incompatibilities in the development of the concurrent engineering tools to be used for the design of the electrical harnesses' installation." He added that "the learning curve for wiring harness changes was too steep during the complex development phase. We have to update and harmonize the 3D design tools and database – and it will take time to do this." In October, 2006 Airbus outlined a comprehensive strategy to rectify the problems and streamline its production processes.

Singapore Airlines now expected to get its first A380 in October 2007, with Qantas and Emirates getting their first aircraft in August 2008. The additional cost to Airbus' parent company European Aeronautic Defence and Space's (EADS) cash flow was given as €6.3 billion ($7.9 billion).

Airbus for many years had been Europe's aerospace success story. The company brought together the finest names in European aviation, but the sales success had taken eyes away from production inefficiencies and political interference. As Streiff moved to streamline Airbus, he faced mounting political and union pressure in France and Germany to work within the status quo to protect jobs.

Airbus wings for the A340-600 arrive from England at final assembly at Toulouse in the giant A300-600ST Beluga. *Airbus*

The shareholding of publicly listed EADS, which owned Airbus, was complex and highly charged politically. In October 2006 major stakes were held by Germany's auto giant DaimlerChrysler and French holding company SOGEADE, with the Spanish company SEPI having a minority stake.

The A380 production problems were a bitter blow to Airbus because the aircraft itself had met all its performance guarantees and was "a delight to fly", said its pilots.

In March 2007 EADS warned that its current structure is "unsustainable and unacceptable," and unveiled details of its Power8 restructuring plan, which included 10,000 job cuts, closing or selling three plants, spinning off three others, outsourcing half of the proposed A350 XWB production and streamlining its business to achieve a 16% increase in manufacturing productivity.

"Our long-term future is at stake if we don't act now," EADS co-CEO and new Airbus CEO Louis Gallois told media at the time. The plan aims to cut annual costs by €2.1 billion ($2.77 billion) from 2010 and produce €5 billion in cumulative savings by 2010.

The wings for the A380 however will come by sea and road. *Airbus*

CHAPTER 13

Dollars and Sense of Aircraft

A commercial aircraft is a vehicle capable of supporting itself aerodynamically and economically at the same time.
- William B. Stout, designer of the Ford Tri-Motor —-

AIRCRAFT OPERATING COSTS

Aircraft operating costs because of their complexity are almost always misunderstood. Differing requirements in range and capacity have dramatic effects on an aircraft design and thus operating costs.

Typical operating costs for the 747-400 per block hour for US airlines in 2000 were $8,158. This is broadly made up of:

- Crew costs - 32%
- Fuel/Oil - 30%
- Maintenance - 14%
- Ownership, insurance etc - 24%

A reduction in any of the elements of crew, fuel or maintenance will have a beneficial effect on the airline's ability to make a profit and pass those gains on to their passengers through lower fares.

As can be seen from Table 13.1, within each category of aircraft the operating costs have decreased through the years as technology and engines have improved. It can also be noted that while larger aircraft have higher operating costs per hour they have lower seat mile costs than smaller regional jets.

Lockheed invested $6 million in 1940 to design the Constellation. *British Airways Museum*

SELECTED AIRCRAFT OPERATING STATISTICS - 2000

	Seats*	Range	Operating costs per hour	Cents seat/ mile	First Flight
High capacity-twin aisle					
DC-10-30	273	6531m	$6388	4.91	1972
Boeing 747-400	369	8393m	$8158	4.37	1988
Boeing 777-200	266	6525m	$4878	3.76	1996
Medium capacity-single aisle					
Boeing 737-200	117	1868m	$2601	6.75	1967
Boeing 757	172	3468m	$3317	4.81	1982
Boeing 737-800	148	3400m	$2255	3.98	1997
Regional jets					
SAAB 340	34	969m	$831	15.52	1983
ERJ-145	50	1734m	$1151	8.63	1995
ERJ-135	37	1962m	$1028	10.36	1998

Seats-Typical seating in US airline service NOT maximum seating.*
These figures are for US/Canadian airlines only and may not represent world wide fleet experience.
All twin engine aircraft except 747(four) DC10 (three)
First Flight: First flight of that particular model
Older models have a higher maintenance burden.

Table 13.1: US airline records

AIRCRAFT PRODUCTIVITY

The 707 (shown) and the DC-8 made all other aircraft obsolete over night. This 707 is owned by Movie Star John Travolta. *Craig Murray*

According to Ray Whitford's comprehensive 19-part series Fundamentals of Airliner Design published in AIR International, aircraft productivity has leapt since the jet era.

In the piston-engine era, the productivity increase of aircraft was comparatively modest, climbing from an index of about zero when the first passengers were carried in the early 1920s to 20 by 1955.

However since 1955, it has soared:

- The introduction of the first 707 saw that index leap to 60.
- The longer-ranged 707-320 had a productivity index of 125.
- The first 747 had a productivity rating of 240.
- The 747-400 had an index of 300.
- Airbus' A380 pushes that index to 390.

The dramatic improvement in the productivity of aircraft is due to a number of factors, including:

- Improved fuel efficiency
- Reduction in aircraft weight/payload ratio
- Reduced number of cockpit crew
- Less frequent engine maintenance procedures
- Greater reliability reducing the number of engine shut-downs and flight delays

Obviously all the aspects of an aircraft's operation are inter-related so let's explain each one separately for the sake of clarity.

FUEL

As explained in earlier chapters, the advances in engine performance are nothing short of staggering. For the jet engine, which is obviously the most relevant today, there has been a 75% reduction in the fuel consumed per mile traveled. This has resulted in the reduction in the take-off weight of an aircraft for a given mission and has enabled aircraft to fly much longer distances.

Boeing committed $16 million in 1952 launching the 707 program. *Boeing Historical Archives*

AIRCRAFT OPERATING COSTS
Year 2000 US cents
per seat mile
Artwork: Juanita Franzi
180
100
90
80
70
60
50
40
30
20
10
5
Junkers Ju 52
DC-3
Constellation
Concorde
Boeing 707
Boeing
747-100
Boeing
747-400
Airbus
A380
Sources: Airbus, Boeing, Douglas & BAe Systems data
1930
1940
1950
1960
1970
1980
1990
2000
2010

The Convair 880 with its four engines had much higher operating costs than the competing Boeing three-engine 727.
San Diego Aerospace Museum

Fuel is the biggest cost factor in the operation of an aircraft. Aircraft and engine manufacturers, and of course airlines, strive to minimize an aircraft's fuel burn and a tiny one per cent improvement can produce sizeable savings. Airbus says that a 1% fuel saving on its four-engine A340-300 in long-range operations can result in savings of more than $105,000 a year per aircraft.

For a direct comparison let's look at Rolls-Royce's RB211-524G, which powers many of today's 747s. This engine has a specific fuel consumption (sfc) of 0.63, half that of the Ghost turbojet of 1952, yet it delivers more than 12 times the power. However, even bigger engines for the 777 have been manufactured by Rolls-Royce and others, with the biggest being the GE90-115B, which develops 115,000lb of thrust and has a sfc of just 0.53lb/pound of thrust/per hour. As can be seen from the Aircraft Operating Costs Graph 13.2, engine advances have had a dramatic effect on the operating costs of aircraft.

AIRCRAFT WEIGHT

The take-off weight of aircraft per passenger has also decreased steadily over the years with the use of more efficient engines requiring less fuel to go longer distances, the use of light-weight composites in the construction of aircraft and with the use of refined aerodynamic principles enabling fewer support mechanisms within the wing.

COCKPIT CREW

Even though aircraft have become bigger, improved technology has reduced the numbers of flight crew. The Boeing 314 Clipper of 1939 required a flight crew of seven with engineers, navigators and radio operators playing a crucial role on board. Even the first 707s in 1959 required a navigator and a radio operator. The navigator had a viewing bubble in the roof to take star sightings.

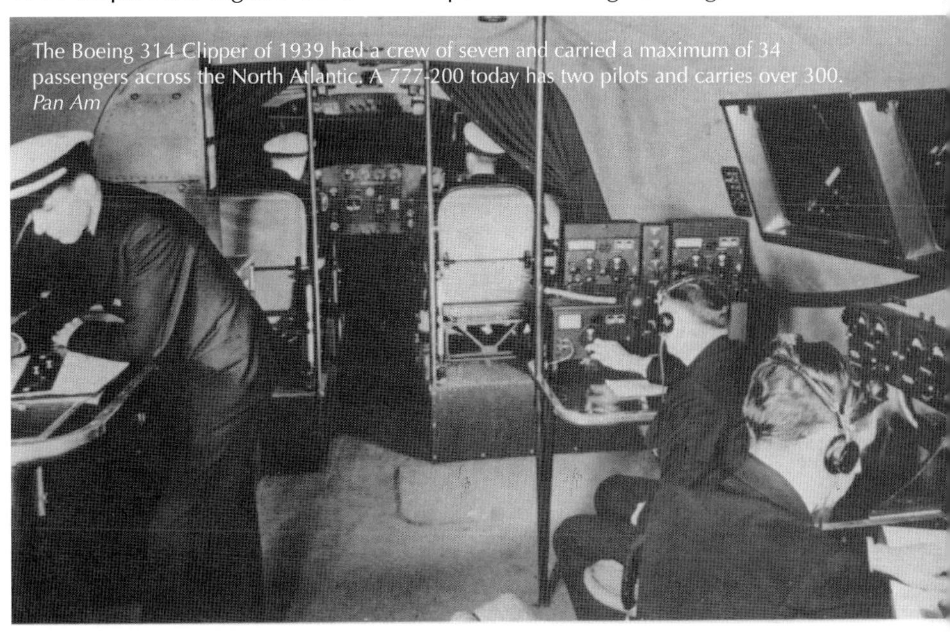
The Boeing 314 Clipper of 1939 had a crew of seven and carried a maximum of 34 passengers across the North Atlantic. A 777-200 today has two pilots and carries over 300.
Pan Am

By the 1990s however, with the introduction of extensive computerization in the cockpit and with the emergence of "glass cockpits" in the A320 (soon followed by the 747-400), cockpit crews had decreased to just two pilots. Some airlines however, such as Australia's Qantas, chose to operate with a second officer on many flights for cruise flying as part of training. On flights over 10 hours, virtually all airlines carry additional crew and the aircraft also have crew rest areas.

Boeing 737-800
Boeing

ENGINE MAINTENANCE

Also impacting on the productivity of aircraft is the engine maintenance burden. In the piston engine era, the DC-3's Wright Cyclone required overhauls every 500 hours, which extended to 1,500 hours for the engines installed on the DC-6 in the 1950s. Improvements in jet engine reliability changed those numbers dramatically, with overhauls every 5,000 hours in the early1960s, which has steadily climbed to overhauls required every 30,000 hours. The savings are massive not only in labor costs but also in "down-time" for the aircraft.

RELIABILITY

The dramatic increase in reliability of the engines has had a multiplier effect on an airline's operations. Rejected take-offs, in-flight shutdowns (IFSD) and flight cancellations all have a considerable escalating effect on an airline's costs. To put that into perspective, Cathay Pacific says that an IFSD may cost it up to 3% of its annual fuel bill for that aircraft. Airbus claims that an IFSD and diversion of a jet can cost well over $1 million. Reduction in these incidences leads to lower costs and lower airfares.

American Airlines 747
Art Brett

The first Boeing 737-100 just prior to its unveiling in 1967. *Boeing Historical Archives*

ADVANCES IN THE 737

There is possibly no better example of how all this engine technology comes together than on the Boeing 737 – the world's best selling jet. The 737-100 first flew in 1967 and could carry 99 passengers over 1,723 miles (2,775km) with a total payload of 26,000lb (11,818kg). The most recent model in widespread airline service, the 737-800, can carry 189 passengers out to 3,791 miles (6,105km), with a total payload of 45,000lb (20,454kg).

Another way to compare the advancement of technology is to look at the Direct Operating Costs (DOC) of the 737. The 737-200 costs $2,600/hr to operate, while the 737-800 costs just $2,255. The larger 737-800 carries an additional 90 passengers in the same configuration layout but costs $345 less per hour to operate. Lower operating costs mean lower fares for passengers.

One vitally important factor when considering costs of operating aircraft is that the older 737-200 carries a heavy maintenance burden to stay flying. If airlines must spend so much money on maintenance of old aircraft they will then not make sufficient profits to afford the more economical aircraft.

A big part of the performance improvement relates to the specific fuel consumption of the different engines used. The 737-100 sported the JT8D-1 which developed 14,000lb of thrust while the 737-800 has CFM-56-7 engines that pack 27,300lb of thrust while burning 20% less fuel.

The Boeing 737-900ER is the largest and latest model of the world's best selling aircraft and took to the air on 5 September 2006. *Boeing*

Air France A340-300 the classic long haul aircraft lands at the aviation enthusiasts haven - St Maartens in the Netherlands Antilles. *Art Brett*

SHORT HAUL VS LONG HAUL

An aircraft's structure and its weight are directly linked to the number of flight cycles the aircraft will perform in its design life. Short-haul aircraft have to tolerate up to 10 times the number of punishing landings than aircraft like the long-haul 747. Therefore the undercarriage and wing need to be strengthened significantly for higher numbers of flights.

Number of flights is also critical for the fuselage structure, which must withstand repeated pressurization cycles. Typically, a 400-seat 747 will perform one flight a day while the 150-seat, short- haul 737 can fly up to 8 sectors a day —- placing much greater strain on the aircraft's structure and undercarriage.

In order to withstand the additional pressures involved in short-haul flights, these aircraft must be built comparatively stronger, thereby increasing the overall weight per passenger significantly. In fact, the empty weight of the long-haul 747 is 602lb (274kg) per passenger in maximum economy configuration compared with 618lb (281kg) for the short-haul 737-800. This is despite the 747 having a much heavier wing structure to hold fuel for long-range flights, plus a massive undercarriage to support the giant jet.

The level of fuel efficiency is also nowhere near as high on short-haul flights as long-haul flights. As a significant proportion of fuel is expended in the first part of a flight climbing to cruising altitude, an aircraft making several short trips a day cannot take advantage of the level of fuel efficiency that a long-haul aircraft traveling far greater distances in one long flight. For example, a 150-seat A320 flying for 3.5 hours from Los Angeles to Atlanta burns 10 tons of fuel for the whole flight but 2 tons is expended in the first 20 minutes. This is further explained at the beginning of Chapter 15. (See Graph 15.1)

This accounts for the higher costs per passenger of running these short-haul regional aircraft. The higher costs of operating the short-haul aircraft also explain the relatively higher airfares on short flights, although there are additional factors which will be detailed in the next chapter.

CHAPTER 14

Airline Costs and Challenges

These days no one can make money
on the goddamn airline business.
The economics represent sheer hell.
— C. R. Smith, President of American Airlines —

Qantas Boeing 737-800 *Misha Popov*

DAMNED COSTLY BUSINESS

It may be hard to believe but the airline industry is actually the world's least profitable. In 2000, the airlines had a good year with a turnover of $328 billion but a net profit of just $3.7 billion or 1.1%. The industry's best year was in 1966, when it made 6.1% net profit, but in the past 30 years, most airline balance sheets have revealed losses of about 1.5% or profits of about the same order.

In fact, from 1947 to 2001 (54 years), the world scheduled airline industry has made a cumulative (and combined) net profit of $27 million.

In comparison, Microsoft made a profit of $7.4 billion in 2001 alone.

In the US, the situation is sickening. According to data provided by the Air Transportation Association (ATA), up to 2001, the US airline industry had made a cumulative profit of just $9.8 billion since 1927 but up to 2005 the cumulative result was a $16 billion loss.

The airline industry is cyclical in nature with a number of profitable years followed by a number of unprofitable years with the general trend of more and more pronounced peaks and troughs. In the year 2000, the industry completed a profitable cycle with a net profit of $3.7 billion (1.1% margin). In 2001, partly due to the effects of September 11, losses totaled $12 billion. While this is a substantial loss, prior to September 11, a $9 billion loss had been predicted. The event of 9/11, combined with the Iraq War and SARS, devastated airlines through the early part of the 21st century and just when airlines recovered from those hits, the fuel price surged. Airlines have shed billions from their costs to meet these massive challenges and the dramatic effect of the cost cutting can be seen from the results in 2005 and 2006.

In August 2006, the International Air Transport Association (IATA), which represents 261 airlines comprising 94% of international scheduled air traffic, announced a revised forecast for 2006 industry losses. It estimated that airlines would lose $1.7 billion, with a fuel bill of $115 billion calculated for an average oil price of $68 per barrel.

Airlines ended 2005 with a $3.2 billion loss including a fuel bill of $91 billion. "We are still in the red but what other industry could add $24 billion to its second largest cost in a year and still improve the bottom line? Efficiency, hard work and a strong revenue environment have all contributed to this amazing result," said Giovanni Bisignani, IATA's Director General and CEO.

THE BURDEN OF COSTS

Without question, one of the most misunderstood aspects of the airline industry is the enormous costs involved in running an airline. Overlaid on this is the complex economics involved in its interrelationship with aircraft. In fact, airlines have enormous infrastructure costs regardless of where their aircraft fly or how far.

This was put into perspective by Cathay Pacific's Chief Operating Officer Tony Tyler when he told industry journal *Air Transport World* in 2005 that during the SARS outbreak the airline "couldn't cut back any further without closing down as so much of an airline's cost structure is set – no matter how many aircraft you operate." Cathay was worst hit by SARS and its traffic crashed from 40,000 passengers a day to just 5,000. The airline responded by cutting 45% of its capacity but managed to maintain the network.

Airline costs are made up of fixed and variable expenses. Fixed expenses generally do not change with the level of activity that the business conducts. Variable expenses however, move up and down with the amount of activity experienced by the airline.

Fixed costs include the administration costs and depreciation or rental/lease costs of the aircraft and other overheads including Information Technology and insurance, all of which account for about 34% of overall expenditure.
With such a large portion of an airline's costs being fixed, it becomes vitally important to maximize the use of the airline's assets (its people and aircraft) and to minimize the amount of revenue per passenger required in order to cover its fixed costs.

One of the largest ticket items is the aircraft, which typically must be paid for regardless of how much they fly. A 747-400 has a list price of $220.75 million, while the cost of a short-haul 189-seat Boeing 737-800 approaches $67 million. Leasing a 747-400 in 2005 cost an airline between $490,000 and $915,000 a month depending on its condition and a 737-800 about $300,000 a month. Lease costs have surged in recent years.

Variable costs are those which change along with a number of key factors, namely:

- the number of passengers (or volume of freight) an airline carries,
- the number of flights the airline operates, and
- the number (and types) of aircraft the airline operates in order to serve its destinations.

The variable expenses can be split between what is commonly referred to as Direct Operating Costs (DOC) and Indirect Operating Costs. DOCs are the costs specifically related to operating the aircraft such as crew, fuel and maintenance.

In the case of the US as a whole, Direct Operating Costs account for 66% of the airline industry's costs. Fixed costs account for 25% of the overall costs and Indirect Operating Costs make up 9%. It is important to note however that the mix of these costs can vary quite substantially among different types of airlines, as well as across different regions of the world.

The aim for any airline is to manipulate the variable expenses to ensure that the fixed costs are fully covered and provide a margin of profit to enable it to expand or invest in new products.

Other expenses are taxes and charges which we will cover on page 172.

A Southwest 737-7H4 in its "Shamu" livery. Southwest has been the great success story of deregulation. *Art Brett*

LOW COST CULTURE CHALLENGE

Traditional full-service airlines, which are bound by strictly enforced and inflexible union pay rates, are at a distinct disadvantage when compared to the more modern low cost operations which have been able to start with "green fields" agreements with younger staff on much lower wages.

In 2002, United Airlines, one of the world's oldest airlines that dates back to the 1920s, had staff costs 65% higher than those of Southwest Airlines which started the low cost revolution in 1973.

However, it should be noted that there are a number of costs that are unique to full service airlines that require administrative complexity. But over the decades, some full service airlines have become bloated with staff that in many cases will not change their culture with the times.

Charles Darwin wrote: "It is not the strongest of the species that survive, nor the most intelligent, but the one most responsive to change." Former British Airways CEO Rod Eddington warned in 2005 that achieving that responsiveness is extremely difficult. "Changing airline culture is like trying to perform an engine change in flight," he said.

While not all legacy airline CEOs have to face as daunting a task as that, the magnitude of reform required to meet the actual or threatened competition from low cost carriers is enormous, and for many airlines seemingly impossible to achieve.

The reasons for the reluctance to change culture are numerous and varied – staff indifference, strong unions and political interference into state-owned airlines' affairs.

Another reason for the resistance to change is that although employees agree change is needed, they don't believe they themselves need to change. For example, at an American Airlines management conference in the late 1990s, all participants were polled on a series of questions. More than 90% responded positively to the proposition that management, colleagues and subordinates needed to change, but 90% responded in the negative to the item "I need to change how I work with people."

According to Nawal Taneja, author of a number of books for airline practitioners, including *Simpli-Flying and Airline Survival Kit,* and an experienced airline business strategy planner, the problem for culture change is that you have to "see death right between the eyes before the impetus to change becomes strong enough."

In 2001 Aer Lingus faced death and there was no ignoring one of the causes – low cost carrier Ryanair. Taneja observed: "At Aer Lingus not only could you see the tails of the

Ryanair jets but you could see its offices from the Aer Lingus offices." This enabled then CEO Willie Walsh to convince his staff that radical surgery was needed. Since late 2001, the airline has stripped 40% from its costs with the departure of 40% of its staff, including 60% of management.

Balancing that however, Taneja said that if workers today are cynical about management philosophies, they have a reason to be. "The failure of companies such as Enron has had a chilling effect on labor-management relations. While executive management enjoy top-up pension plans, huge bonuses and are seemingly isolated from the effects of bad decisions, the rank-and-file have felt pain with pressure for improved productivity and in some cases unfunded pension plans."

Author Kaye Shackford agreed with Taneja. In her book *Charting a Wiser Course: How Aviation Can Address the Human Side of Change*, she maintained "you can't treat your shareholders better than your partners, customers or employees."

For many airline staff of legacy carriers, such as British Airways and United Airlines, the success of the new low cost carriers is hard to judge because they often don't operate into major traditional airports such as Heathrow or Chicago's O'Hare. In the United Kingdom, low cost success stories Ryanair and Easyjet use Stansted and Luton airports.

One of the major problems in changing staff culture is how to strip out complexity. According to Taneja, airlines have added enormous complexity to their operations and processes over time and staff are ingrained with these systems. "Enormous fleet and network complexity and labor contracts have been built around these complex systems and processes which often relate to charging higher and more complex fares. Unfortunately, many passengers now are not willing to pay for this complexity given the Web-enabled transparency in fares and the increasing availability of low-fare services by the LCCs."

Yet another view on culture change comes from one operations manager, who explained that the enormous emphasis on safety in airline culture is built around strict management systems and attention to detail "which is the opposite of flexibility and flair." He noted that "airlines have built up a rigid model over decades for very sound reasons." Another obstacle in the race to convince management and staff of the need for culture change is the marketing by airlines of the luxury side of travel. "The network and legacy airlines have created the image of luxury and excess to attract high-yield passengers," noted one executive. "This contradicts the message of restraint and cutbacks we are preaching to our staff."

One airline that has demonstrated the ability to create a winning culture for more than three decades is Southwest Airlines. That performance was built around total staff involvement capped by profit-sharing and one of the most generous staff stock option bonus systems in the industry. In the industry, it is a well known fact that "it's harder to get a job at Southwest than getting into Harvard Business School."

United A320
Airbus

Up to March 2007, Ryanair had ordered 281 Boeing 737-800s. *Boeing*

WHAT'S A FRILL?

As legacy airlines struggle to reduce costs to combat the rising tide of low cost carriers (LCCs), those airlines have armed themselves with the legacy carriers' last weapon in their armory – frills.

New York's JetBlue led the way with seatback videos, Westjet in Canada recently introduced free TV and Australia's Virgin Blue and Jetstar have added satellite TV or DVD players.

But for many passengers, there is more to the definition of a frill than a free meal or a seatback video. It can be ease of booking or quick check-in, getting a genuine smile from obliging staff or seamless baggage retrieval at the end of a flight, which if not a frill, is certainly a "thrill" helping to bring passengers back time and again.

And the frills vary from one passenger group to another and from one country to another. For the British holidaymaker, it is a bargain fare. For a German businessman, it is being able to pay to jump a queue. And to an Aussie, it is being able to watch live sport on a seatback video screen.

The strategies for most LCCs are two-fold – to take on the legacy airlines and attract higher yield passengers and to add points of difference from other LCCs which are now encroaching on each other's turf. In Europe, in 2005 there were 78 airlines of the LCC species and of the 115 routes where LCCs had competition, 80 involved another LCC.

To meet the competitive challenge, LCCs like Ryanair have relentlessly stripped costs out of their operations but have added value and online features. Ryanair generated up to 16% of its revenue from sources other than the ticket price. Ryanair's CEO Michael O'Leary said he believes the airline industry is like the cinema industry with more revenue being generated through ancillary sales than tickets. In 2003 Ryanair gave 20% of its seats away at zero euros and by 2008 it aims for 50% free seats.

But in 2003, Ryanair's commission on car rentals was $36.7 million, in-flight sales accounted for $30.17 million, internet revenue was $15.7 million and non-flight revenue was $49.9 million. At the same time, the airline constantly topped the European on-time statistics, had fewer lost bags and fewer cancellations. For many passengers, that is a frill.

In Germany, LCCs such as Germanwings and Air Berlin have responded to market demands by adding some targeted frills. Germanwings has blended corporate travelers' demand for recognition of their frequent flying with online flexibility including special customer ID log-in to its website for bookings.

Over 40% of Germanwings passengers were business travelers and the airline had over 500 companies signed up with contracts. Like Ryanair, Germanwings also targeted ancillary commission revenue with links to hotel chains and car rentals and promoted special accommodation and car deals. On board, Germanwings returned to the strict LCC model charging for snacks, wine, beer and even coffee.

Air Berlin bucked the LCC model when it comes to frills. It offered a comprehensive frequent flyer program with partners, such as Hertz, American Express and Radisson, served hot meals on flights over 3½ hours and launched a City Shuttle in 2002 to connect German airports with major European cities.

In the US, the opposite appears to be the case, said Washington-based airline analyst George Hamlin. "Some legacy carriers have performed the extraordinary feat of reducing service levels below the offerings of the LCCs," he said.

Hamlin lauded airlines like JetBlue, which "deliver great service, plenty of legroom and a seatback video. What more could you want? On JetBlue and Southwest, you feel as if you are in control and as if you are getting value ... and that is a frill – and a thrill."

Hamlin was scathing about those legacy airlines that had eliminated minor frills to save just $2 million a year. "They are nibbling around the edges to save peanuts," he claimed. And while some airlines strive to save money by charging $1 for pretzels and nuts, JetBlue added 100 channels to its XM Satellite Radio to complement its 36 free DIRECTV channels. "All of a sudden some legacy carriers are the no-frills airlines," said Hamlin.

US Air Travel Consumer Reports added an interesting perspective on what matters to travelers. Flight problems (28.1%), Baggage (26.7%), Customer Service (10.6%) and Reservation/Ticketing and Boarding (11.7%) topped the list, while complaints about fares barely made the radar at 2.6%. And it came as no surprise to find that the airlines that were making profits – Southwest and JetBlue – had the lowest level of complaints, right across all areas.

In Asia, the dynamics are different but the need for frills to meet varying market needs is the same. For airlines, such as Kuala Lumpur-based AirAsia, the frills or thrills they offer Malaysians are the ability to fly for the first time – so in-flight frills mean little. Obviously, as the market matures, the airline is likely to look at value-add to meet expanding market demands.

LONG HAUL LOW COST

With the runaway success of low cost carriers (LCCs) on short-haul routes, existing and potential carriers have looked at long haul low cost (LHLC) as the next battleground – but the battle had already started. A closer look will show that that battle has always been going on – but under a different name.

JetStar International, Qantas' low cost long haul subsidiary will use the 787 to operate routes from Australia to Asia, the US and Europe. *Boeing*

On pages 102 to 105, we followed the great stories of Trippe, Laker and Branson, who each pioneered lower fares from the 1950s through to the 1990s by introducing economy class and then mass travel with 747s and DC-10s and more recently by providing passengers with fun and cheeky competitions with lower overheads.

But the structural and competitive hurdles to bringing everyday low-fare competition to intercon-tinental markets on a scheduled basis are daunting. Among them is the difficulty, and in some cases, the impossibility of gaining attractively timed operating slots and real estate at international gateways. And many governments are in no hurry to encourage low-fare competition if it will risk their own investment in a state-owned or controlled airline.

The very nature of long-haul operations reduces some of the advantages enjoyed by LCCs. For example, faster ground turns can make the difference between eight and nine trips on a 300 mile segment but mean a lot less when the plane is going to be airborne for 8 to 9 hours.

"Flying long-haul doesn't offer the same cost advantages over legacy carriers, and the successful startup service requires an enormous venture capital commitment for high-ticket items such as long-range aircraft, fleet recon-figurations and infrastructure," suggested William Franke, former chairman of America West Airlines and managing partner of Indigo Partners LLC, which owned a share in Singapore-based LCC Tiger Airways.

Former Cathay Pacific and British Airways chief Sir Rod Eddington told trade journal *Air Transport World* in 2004 that one of the biggest challenges to a long-haul LCC is "the back end of a 747" because of the large volume of seats that can be discounted. A similar view was held at Cathay Pacific, which used both its 747s and 777-300s to offer very cheap fares. In 2005 Cathay Pacific countered Singapore LCC Tiger Airways 160-seat A320s with a conditional HK$930 ($119) return fare between Hong Kong and Singapore.

However, British Airways and Cathay Pacific are well-run, lean operations that offered cheap fares and still managed to make profits. There were many that were not and they may be the target of a new breed of long-haul low cost airlines.

As we saw on page 152, some legacy carriers, such as Ireland's Aer Lingus, have reinvented themselves. Aer Lingus applied the low cost ruler across its entire operation – short haul to Europe and long haul to the US and in 2005 offered fares across the Atlantic from $422 with no restrictions and no minimum stays. They upheld the philosophy: "If the seat is there, you get it." Another who had eyed the LHLC proposition was Dubai based Emirates, which wanted to use its eventual fleet of giant A380s to operate a long-haul low cost operation. The airline's President Tim Clark speaking at the unveiling of the A380 in January 2005 told media that he saw a market for an all-economy, 780-seat configuration operating from low cost airports like Stansted in the UK to Adelaide in Australia via Colombo in Sri Lanka at a fare of $520 return and a breakeven point at 80% load factor. He predicted that flights from Stansted to Burbank, California would cost $390.

In an ironic twist, Clark, who presided over an airline that was at the cutting edge of service delivery and appointments, believed he was well-qualified to outline what an LHLC carrier should offer. "The A380 would only have beverage stations and passengers would bring their own catering requirements. Video-on-demand and all drinks would be charged for and passengers limited to one piece of baggage at 25kg."

At the other end of the size spectrum, others saw the 230 to 400 seat Boeing 787 as the benchmark for LHLC operations. According to Mike Bair, VP and GM-787 Program, the 787 would be "sensational for a long-haul operation, offering similar seat-mile costs to the A380 with much lower trip costs."

And many lower cost airlines have lined up to fly the 787 such as Australia's Qantas LHLC subsidiary JetStar International which will get its first 787 in 2008. That airline had taken over from Qantas many of the routes that were marginal or loss makers.

Cathay Pacific uses the superb economics of its 385-seat 777-300s to combat low cost carriers.

NOT JUST BUMS ON SEATS

Most airline revenue comes from passengers, although for many airlines, an increasingly important component is cargo.

EVA Air Boeing 777-300ER touches down at Farnborough in July 2006.

Aircraft such as the A340-600 and the 777-300 have more under-floor cargo capability than a 707 pure freighter. In the US domestic airline system, 75% of revenue came from passengers and 15% from cargo shippers. But airlines such as Hong Kong's Cathay Pacific earned 30% of their revenue from cargo, while EVA Air in Taiwan gleaned 45% of its revenue from cargo.

In fact, many airlines have planned their flights around such cargo as live lobsters. Cathay Pacific scheduled its four times weekly Perth, Western Australia to Hong Kong flights around getting lobsters to the markets in Taipei and Tokyo at the right time. And the right time is just before the lobsters – which have been put to sleep over 12 hours in progressively colder water – wake up. If the lobsters wake before they are unpacked they fret and shed their legs making them un-saleable.

Airlines have used the power of the internet to not only cut out the cost of distribution but also to sell other products, such as hire cars, holiday packages and hotels.

As we saw on page 154, more and more airlines are selling products in-flight such as access to seatback TV, food, drinks, books and even the right to jump the queue. Some airlines are also selling exit row seats at a slight premium.

FLEET SIMPLIFICATION

A major cost for airlines is the outlay required to keep a mixture of aircraft types in a fleet to perform very different missions both in capacity and range.

For instance, up until 1996, airlines that wanted to fly long distance non-stop routes such as Singapore to Europe had one choice – a 416-seat Boeing 747. For many airlines, the 747-400 was much too big and costly and some airlines had only a handful of 747s just to stay competitive.

It was only in the mid 1990s, thanks to significant improvements in engine technology, that the 300-seat Airbus A340 and Boeing's 777-200ER emerged permitting flights of similar range to the 747 and giving airlines lower capacity and cost options. It is also only in recent times that medium capacity airliners – like the 180-seat 737 – have been able to fly transcontinental across the United States in all weather.

To cover its widely varying markets, Australia's Qantas operated a mixed fleet of 747s, A330s, 767s and 737s, while its subsidiaries and regional partners operated A320s, 717s, BAe146s and Dash 8s. In the US, American Airlines had a similar fleet mix with 777s, 767s, 757s, 737s, MD-80s and A300B4s and its American Eagle regional operator had CRJ700s, ERJ-135/140/145s and Saab 340Bs.

And within that mix of aircraft, there were different sized variants and vastly different configurations depending on whether the aircraft were used for international or domestic configurations and what market they served.

A good example of this would be Qantas which had three configurations for its 747-400s. The highest capacity configuration had 56 business and 356 economy seats for routes to Frankfurt, while the least was 14 first, 64 business and 265 economy for US routes. Its 747-300s had two different two-class configurations.

In stark contrast, airlines such as Southwest Airlines and Australia's Virgin Blue have operated just one type, the Boeing 737. By operating one type, airlines eliminate problems with crew rosters, are able to simplify maintenance, reduce staff training costs and slash spares holdings.

Boeing's new 787, which fliesin August 2007, brings with it an entirely new dimension to airline flexibility. The aircraft is what is termed mid-sized, seating between 220 and 350 depending on model and configuration, but it can fly ranges up to 8,801nm (16,299km) economically.

For the first time, airlines will be able to connect non-stop literally hundreds of city pairs across the globe cost-effectively.

Australia's leading LCC, Virgin Blue has focused on the 180-seat Boeing 737-800 and -700 for its fleet operating over 50. However, the airline has now moved to buy smaller 100-seat E-jets for regional routes. *Misha Popov*

STAFF COSTS

Being a service industry, airlines – even low cost carriers (LCCs) – are labor intensive. Each airline employs armies of pilots, flight attendants, mechanics, baggage handlers, reservation agents, check-in staff, security personnel, catering staff, cleaners, administration staff, accountants, lawyers – and the list goes on.

However, the actual numbers of airline staff can be significantly distorted because many airlines subcontract a number of functions, such as ground handling, catering, maintenance, information technology and even flight crew services.

The cost of staffing an airline is also related to the region in which it operates and the level of wages in that region, whether it is a short-haul or long-haul airline operation and whether it is a full-service or a "no frills" operation.

Qantas' staff costs in the 2005-06 year were 26% of its total expenses – the highest in the South-East Asian region. (see Table 14.1)

Qantas Costs 2005-06

	COSTS	PERCENTAGE
Staff	$3,321,700,000	26%
Marketing	$469,600,000	4%
Aircraft operating	$2,525,300,000	19%
Fuel and Oil	$2,802,300,000	22%
Property	$320,000,000	2%
Computer/communication	$487,500,000	4%
Depreciation	$1,249,800,000	10%
Operating Lease rentals	$355,700,000	3%
Tours/Travel costs of sales	$591,200,000	5%
Capacity Hire Insurance	$369,600,000	3%
Other	$467,000,000	4%
	$12,959,700,000	100%

Figures in Australian Dollars

Table 14.1 Source: Qantas Annual Report 2005/06

By contrast, airlines such as Singapore Airlines and Hong Kong-based Cathay Pacific are purely international, although they do have some short-haul international routes. In both cases, their labor costs are approximately 21% of total costs – much lower than Qantas's.

A factor in the lower staff costs for these airlines is the nature of long-haul operations which tend to be less labor-intensive. In addition, both Asian-based airlines enjoy overall better productivity from their staff and they employ leaner work practices.

American Airlines, the world's biggest airline and considered by many as the most successful of the past century, was tragically affected by the events of September 11 and has also felt the impact of the upsurge in low cost airlines. However, over the past few years, it has staged a remarkable turnaround and is now in profit. Table 14.2 lists its costs for the year 2005.

These figures reflect the generally higher staff costs for many US full-service airlines which in some efficiency gains but also because of the significantly higher cost of fuel.

The high labor costs are a legacy of heavily unionized workforces combined with short-haul operations. Airlines such as Pan Am, Eastern and TWA were badly affected by high labor costs, which was a significant factor in their eventual demise, while United Airlines, US Air, Delta Airlines and Northwest Airlines sought Chapter 11 protection after 9/11.

In 2005, the cost of running American Airlines was $20.8

billion and the airline flew 222.6 billion Revenue Passenger Kilometers (RPKs) – that is, passengers x kilometers flown – for a cost of $0.09/RPK. Southwest Airlines, however, with its one aircraft type and a more simplified operation, produced impressive figures for 2005 amounting to just $0.069/RPK.

The fact remains that costs per employee are among the highest of any industry, according to the US Air Transport Association (ATA). The ATA, which counts in its members the nine largest US airlines, says the average wage and benefits paid to the 569,778 employees of its members was $78,636, 2005.

That figure should be balanced with the US national average earnings at $16.79 an hour for a 39.6-hour week – or $34,573.96 a year. Employees in the leisure and hospitality industry were paid just $9.49 an hour. (see table 14.3)

One of the starkest examples of the staff cost efficiencies of low cost airlines is the comparison between the defunct Ansett Australia which collapsed in late 2001 and Australian LCC Virgin Blue's passenger uplift/staff ratio. According to Virgin Blue figures, the airline carried more passengers than Ansett on major Australian domestic trunk routes for the year ending 31 March 2004, with only a third of the staff. The two airlines' RPM/RPK figures were almost identical at this time, however Virgin Blue only has 3,300 staff compared with the 10,000 staff that Ansett employed for its domestic trunk route division.

American Airlines 2005

	COSTS	PERCENTAGE
Staff	$6,755,000,000	32%
Commissions	$1,113,000,000	5%
Aircraft operating	$2,809,000,000	14%
Fuel and Oil	$5,615,000,000	27%
Maintenance	$989,000,000	5%
Depreciation	$1,164,000,000	6%
Operating Lease rentals	$591,000,000	3%
Other rentals and landing fees	$1,262,000,000	6%
Food service	$507,000,000	2%
Other		0%
	$20,805,000,000	100%

Table 14.2 Source: American Airlines Annual Report 2005

US Department of Labour

OCCUPATION	YEARLY INCOME
Airline pilots, copilots, and flight engineers	$137,160
Aircraft mechanics and service technicians	$54,890
First-line supervisors/managers of office and administrative support workers	$47,450
Flight attendants	$43,470
Baggage porters and bellhops	$38,600
Transportation workers, all other	$37,790
Cargo and freight agents	$36,700
Reservation and transportation ticket agents and travel clerks	$31,450
Customer service representatives	$28,420
Laborers and freight, stock, and material movers	$21,570

Table 14.3 Source U.S. Department of Labor

COMPUTERS

Another way in which technology has influenced the airline industry (besides improving efficiency of aircraft and manufacturing techniques) is through the use of computers for reservations. Load factors for airlines were stuck at 65% for years, but with the advent of computers and the strategy of "yield management", load factors have increased to 75%.

In November 2005, Emirates ordered 42 more 777s to add to its fleet of 51 777 aircraft. The deal was worth $9.7 billion at list prices and included 24 Boeing 777-300ERs, 10 777-200LRs and 8 777-200Fs. *Boeing*

Airlines recognize that once aircraft depart, the return from any unoccupied seats or unused cargo space evaporates. It is crucially important to sell as much of the space on board as possible at the highest possible price. How airlines can do this, without pricing themselves out of the market or taking off below breakeven point, is something of a science. This is addressed by highly sophisticated computer systems.

Yield management enables airlines to smooth out demand. Typically, a flight will be in an airline's computer system for 12 months and the fares to be offered, including frequent flyer seats, will be set up using historical data.

During the next 12 months, the computer system will interrogate the flight continuously to determine which fare categories are being sold. When one fare level is sold out, the system may take unsold seats from another fare level and re-price them.

Yield or revenue management systems have served the airlines and the public well. Some claim that implementing a yield management system can increase revenue by 8-10%. It is important to note however that the systems have been unable to handle unpredictable passenger behavior associated with such global events as terrorist attacks and SARS, so there were considerably more manual inputs in the year after September 11.

The computer generated system of yield management also helps manage such anomalies as "no shows" through overbooking. It can also smooth out the effects of seasonal variations.

Overbooking is not so much of a problem on contract charter flights like those operated by National Jet Systems to resource sites in Western Australia.

OVERBOOKING AND SEASONAL VARIATIONS

An often misunderstood aspect of airline operations outside the US is overbooking. Airlines overbook flights to compensate for the often large number of cancellations or no-shows.

The no-shows are almost always business travelers who pay top price for a fully refundable ticket but when they don't show up for their flight, their empty seat is wasted. That empty seat represents an additional opportunity for the airline to increase revenue and satisfy last-minute demand. The practice is controlled once again by careful computer analysis of historical demand for a flight, together with the elements of economic principles and human behavior.

The problem is that many business travelers make multiple bookings for flights over say a three-hour time frame, usually later in the day to ensure they catch at least one flight home after a business meeting. On some flights, such as Friday afternoon/evening New York-Chicago, overbooking can be 10% or higher.

However, when the calculations and assumptions are wrong and more people show up for a flight than there are seats available, airlines will offer incentives to get people to give up their seats. Such things as free tickets and first-class upgrades are offered. In most cases, there is a scramble to take the free tickets, upgrades or frequent flyer points. But when there are not enough volunteers, airlines will deny boarding to the last passenger to check in. Overbooking typically occurs far less often on international flights or on flights where there may be only one trip a day.

The airline business is seasonal. The summer months are extremely busy, with travelers taking vacations or just taking advantage of fine weather. Winter, on the other hand, is slower with the weather making travel in some places quite difficult.

The exceptions to this, of course, are national family holidays, such as Thanksgiving, religious and New Year breaks. Another seasonal factor is the post-Christmas lull which continues to affect many airlines around the world. Both corporate and personal travel is influenced with travelers spending heavily in the lead-up to December and cutting back into the next year.

In the past, the result of the natural peaks and troughs in travel patterns was that airline revenues rose and fell significantly through the course of the year. While this pattern continues to some extent today, through sophisticated yield management, airlines have successfully evened out the peaks in demand by incentive pricing.

Qantas' first 737-838 was delivered on 14 January 2002. The aircraft was the third to sport a tribute livery to the Aboriginal culture. Dubbed "Yananyi Dreaming", the design reflected images of the Western Desert landscape surrounding Uluru (Ayres Rock). *Qantas*

AIRLINE PRODUCTIVITY SOARS

Looking at the airline industry as a whole, productivity has always been on the increase but never more dramatically than since 1945. According to the International Air Transport Association (IATA), airline productivity has leapt by just over 1,000% to 310,475 Freight Tonne Kilometers (FTKs) per employee in the period from 1945 to 1999 (that is, the total weight of passengers and cargo divided by employees and the distance traveled in kilometers).

According to International Civil Aviation Organization (ICAO), since 1970, aircraft, fuel and labor trends have been positive overall and have improved each year by:

- Aircraft productivity 3.2%
- Fuel productivity 2.6%
- Labor productivity 6.5%

The combined effect is an annual 5% productivity gain. Although labor productivity shows the biggest percentage gain, ICAO claims that much of that is the result of more efficient and bigger aircraft. As Chapter 12 explained, aircraft productivity has soared through engine technology, while higher utilization and load factors have leapt through computer based yield management programs.

A very important aspect of airline productivity is the ability of airlines to increase aircraft utilization – the number of hours an aircraft flies. In the past 20 years, utilization has risen in all regions with the highest gains in the US at 42%. According to Rolls-Royce, average daily utilization of aircraft in the US with mature utilization/yield management software is nearly 12 hours while Third World airlines struggle with rates of seven hours.

If the ownership costs and crew costs can be spread across more operating hours, the hourly cost of operation will reduce. Successful airlines manage to get maximum utilization out of the aircraft and the best from their people in order to achieve the goal of profitability.

This is often realized by flying longer distances, "turning around" the aircraft as quickly as possible when it is on the ground and flying the aircraft longer and/or more frequently per day.

Flying the aircraft for more hours is often achieved by flying at night from airports without noise curfews to land at noise curfew airports after 6am. Better utilization of its assets is one natural benefit to long-haul airlines whereas it is much harder to achieve for short-haul airlines.

These productivity gains through massive advances in technology are largely responsible for keeping air fares static relating to average weekly earnings as can be seen in Graph 14.9 (opposite).

How cheap it is to soar...

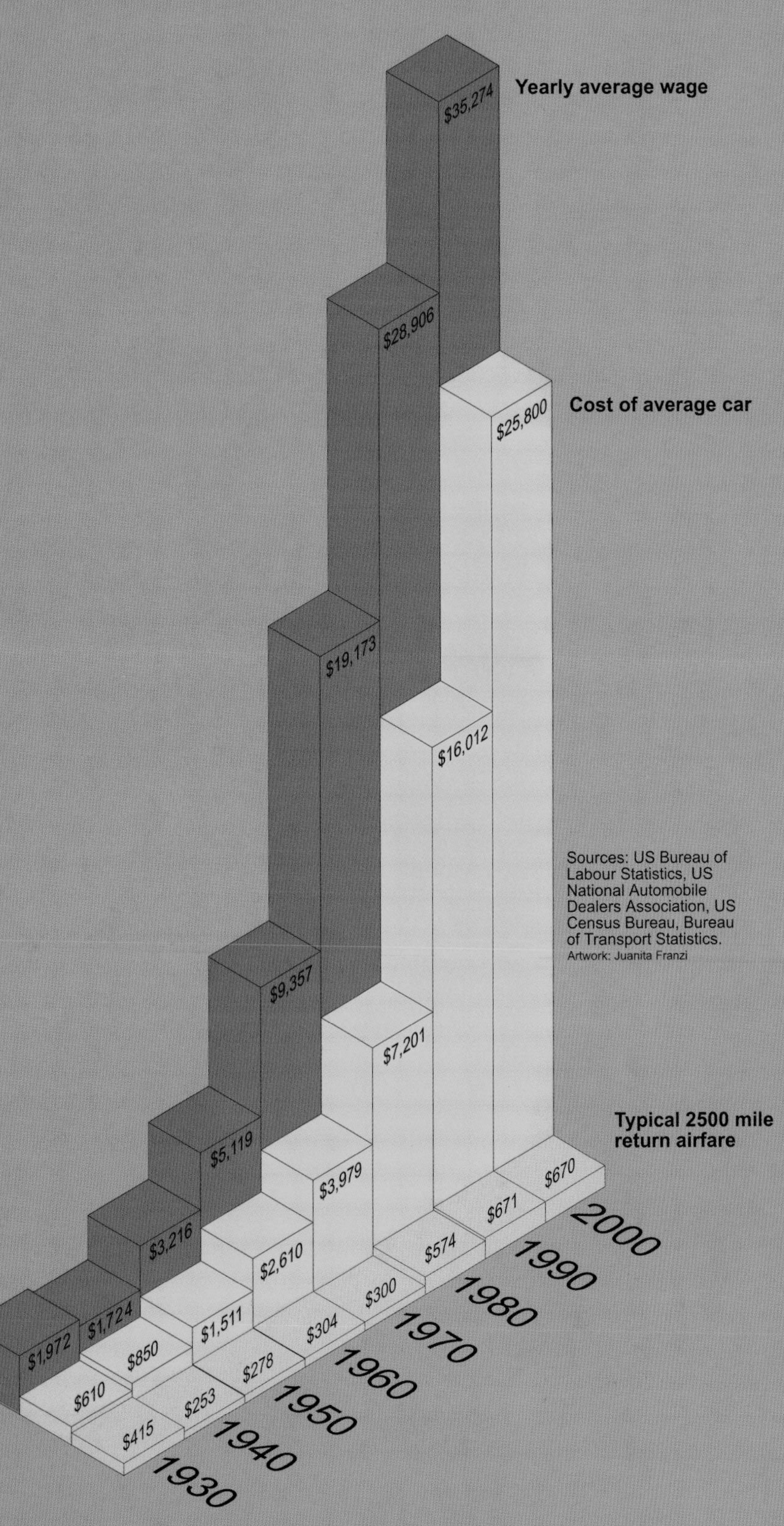

Graph 14.9

Air traffic congestion is a massive cost for the industry. This is a line up at Boston's Logan airport.
Mark Garfinkel

REDUCING DELAYS

In order to ensure that an airline utilizes its assets to the maximum it is absolutely essential that aircraft departure delays are reduced to the absolute minimum. They constitute a "hidden" cost as far as the public is concerned and add enormous costs to air traffic control (ATC).

When a delay occurs it creates a series of chain-reactions that affect crew rosters, re-allocation of seats for passengers, even complicating planned maintenance activity as the aircraft might not finish in the port it was originally destined for, with the subsequent effect of an increase in costs.

Delays also lead to:

- extra fuel burnt while aircraft wait to take-off
- increased overtime payments to staff
- increased displacement expenses as a result of the crew "over-nighting" away from their home-base

According to a Campbell Hill Aviation Group study, aircraft congestion due to ATC delays in the US in 2000, cost the country $9.40 billion. The analysis found that delays cost passengers $4.70 billion. Putting that into another perspective delays in 2000 were the equivalent of having over 400 jets grounded for an entire year!

The study also found that even if all the aviation infrastructure costs outlined by the Federal Aviation Authority (FAA) were implemented, there would be even more delays in 2012. And the cumulative costs to the US economy by 2012 would reach $158 billion.

The story in Europe is no better.

EuroControl estimates put the cost to passengers and airlines of ATC delays in Europe at $5.50 billion a year. In 1999 25% of flights experienced an ATC delay of more than 21 minutes as the result of the underlying absence of a uniform, coordinated European air traffic control system.

CROSS SUBSIDIZATION

When considering new destinations, airline management can apply the principle of "cross-subsidization". The term cross-subsidization is sometimes mistakenly applied to an airline using profits from one route to balance losses on another. It actually refers to an airline applying the infrastructure costs, which are underpinned by its profitable network, across all its operations, allowing it to operate flights to marginal destinations without applying the full costs of operation to that route.

This can occur for instance, when an aircraft has an overnight stop, and the airline uses it during that time - often referred to as "back-of-the-clock" - for tourist flights or charters. An airline may also purchase aircraft to add to its fleet to serve marginal routes, perhaps linking several towns in a "milk run". With maintenance, spares, training and market infrastructure already in place, the employment of additional aircraft is not as much of a cost burden as it would otherwise be. In this way, the marginal routes are said to be cross-subsidized by the entire operation.

Since it was launched in 1984 Airbus has sold 5,053 of the A320 family.
Airbus

Cathay Pacific has some of the world's longest air routes. Here on a "shorter sector" a 747-400 heads home to Hong Kong from Sydney. *Craig Murray*

COMPARING REGIONS

Another major factor affecting the profitability of an airline is the region and/or country in which it is based.

Costs can vary quite considerably between countries due to such differences as:

- labor practices
- social requirements
- infrastructure costs
- taxes
- industry practices

As industries around the world become more global in nature, these differences become increasingly evident.

Airlines based in Southeast Asia enjoy several benefits from their geographic location. These relate to:

- Lower standard wage rates compared to airlines outside this region
- Lower labour costs as a result of a higher level of productivity
- The greater distance between Southeast Asia and other parts of the world. Due to the longer distances between destinations, their level of long-haul operations allows for higher aircraft utilisation.

One airline facing a significant challenge due to the geographic differences between regions is Australia's Qantas.

Although Qantas benefits from longer-haul operations between Australia and the United States as well as Australia and Europe, it also has a significant number of regional short-haul operations. The work practices in Southeast Asia give the airlines in this region (such as Singapore Airlines, Malaysia Airlines and Cathay Pacific) a labor cost advantage that is very difficult for Qantas to match. This prompted Qantas to set up crew bases in Southeast Asia and New Zealand in order to bring its costs more in line with those of its competitors.

The US is at an even greater disadvantage with staff costs accounting for 41% of operating costs. Until recently, US airlines have borne the brunt of a demanding and highly unionized workforce, adversely affecting their profitability. As a result, a number of new-generation "Low Cost, No Frills" Airlines have emerged which have been able to avoid legacy costs, keep their operations simple and thereby improve their level of profitability.

China Airlines A340-300
Airbus

CHAPTER 15

Decline in Airfares

This is the most important aviation development since Lindbergh's flight. In one fell swoop, we have shrunken the earth.
– Juan Trippe, founder of Pan Am on the introduction of jet aircraft and its downward effect on airfares –

WHAT MAKES A FARE FAIR?

Rarely has a product or service received so much comment in respect of its price as airline fares. Perhaps the growing importance of air travel in all of our lives has put airfares in an ever sharper focus.

The world has become a global market for businesses, and we all want to see the far corners of the world to enjoy our leisure time as television and the internet have brought these places into our daily lives.

There are a host of costs that make up an airline fare. These include the cost of operating the aircraft, the cost of selling the ticket, the cost of administration, the cost of engineering personnel and parts, along with a host of government fees, navigation and airport taxes.

According to US Department of Transport, US airlines' direct and fixed operating costs are made up of:

- Flying Operations 33.7%
- Maintenance 10.7%
- Aircraft and Traffic Service 15.3%
- Passenger Service 7.2%
- General & Administrative 6.5%
- Promotion & Sales 6.5%
- Transport Related 15.0%
- Depreciation & Amortization 5.1%

It can be seen from these figures that a significant portion of the costs are fixed. These costs apply regardless of the distance of a particular flight.

This is a significant area of mis-understanding for the public when they try to fathom the wide variations in fares between short-haul and long-haul flights. Aircraft are designed for very specific missions from short haul to long haul with flight times varying from about one hour to 18 hours for the A340-500 or the 777-200LR. The effect of the short-haul mission on the structure of the aircraft is dealt with in Chapter 13, as is the role of engines and fuel consumption.

Let's examine these fixed costs more closely and assess how they affect the cost of an airline ticket. Australia's Qantas is a good example. In 2005/06, the airline's expenditure topped A$12.95 billion and its costs were made up of:

- Fuel & Oil 22%
- Aircraft Operating Costs 19%
- Depreciation 10%
- Selling and Marketing 9%
- Staff 26%
- Other Expenses 14%

It can been seen from these figures that selling/marketing (9%) and Other (14%) account for 23% of costs and are largely unrelated to aircraft, while at least half of the staff costs can be considered non-flying, making a total of 36%. So whether the flight travels 1,000 miles or 10,000 miles, at least 36% of the costs are fixed. When you then overlay the high cost of getting an aircraft to cruise altitude relative to the total flight, plus the one-off taxes and airport charges which we cover on page 172, the basic cost or flag-fall cost, may be as high as 60% on some short haul routes. The dramatic effect of these one-off costs can be seen in Graph 15.1.

Early morning lineup at Denver International Airport . *Tim Samples*

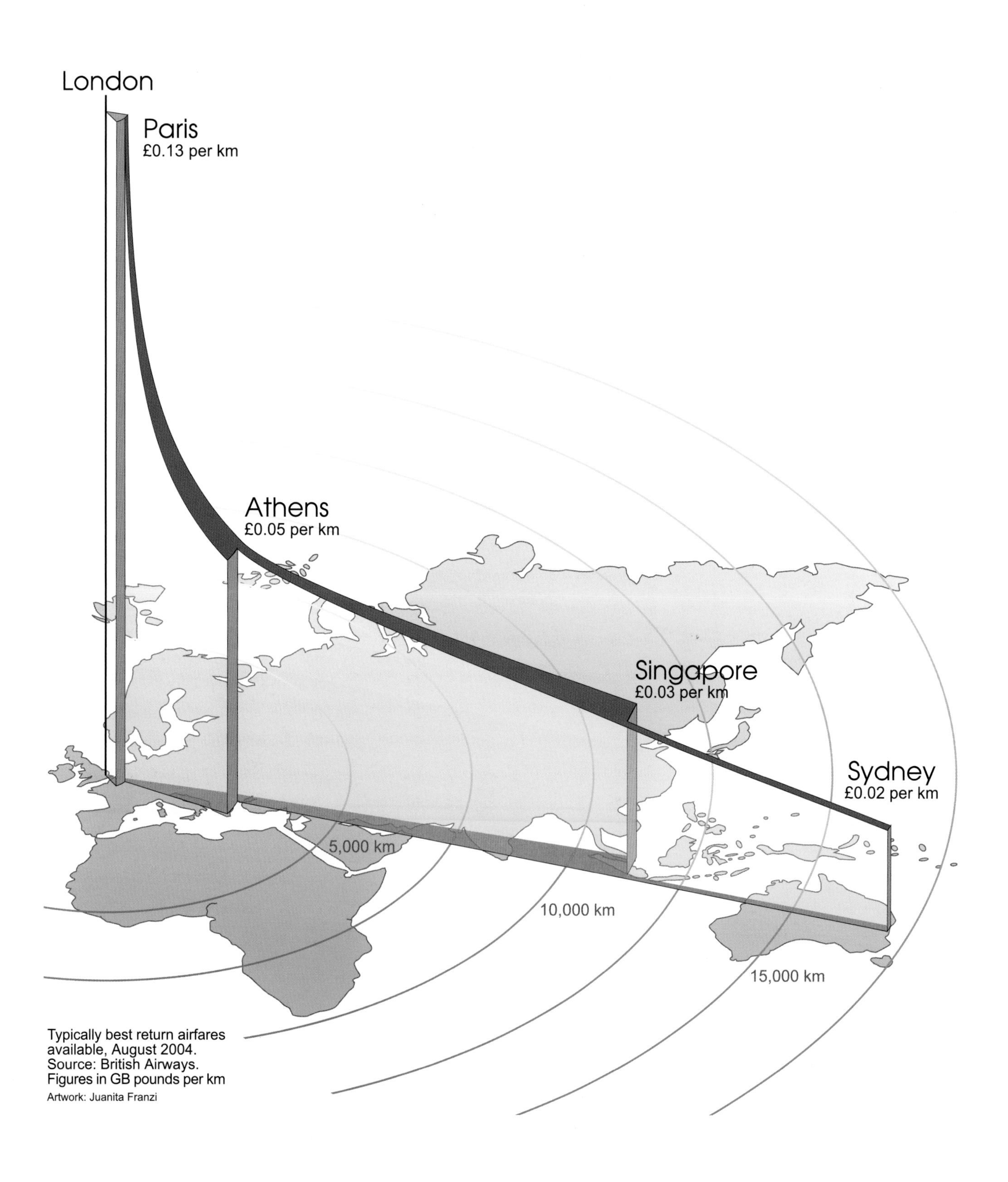
Effect of Distance on Airfares
London
Paris
£0.13 per km
Athens
£0.05 per km
Singapore
£0.03 per km
Sydney
£0.02 per km
5,000 km
10,000 km
15,000 km
Typically best return airfares available, August 2004.
Source: British Airways.
Figures in GB pounds per km
Artwork: Juanita Franzi

Graph 15.1

An Avro Anson which can carry six passengers – identical to Gulf Air's first aircraft – poses with a factory fresh A330-200 which can carry 406.

Also fixed are the range of taxes and charges that are applied to the cost of an airline ticket. An example of this is the return airfare from Sydney to Toronto with Air Canada which cost A$2,521.48 in September 2006 but the charges and taxes push that fare up to A$3,019.62.

These extra costs are made up of:

• Navcan and Surcharges	$36.52
• Fuel Surcharge	$290.00
• Canada Airport Improvement Fee	$18.20
• U.S Agriculture Fee	$6.70
• Canada Security Charge	$20.60
• U.S Passenger Facility Charge	$6.10
• Australia Passenger Service Charge	$41.92
• Australia Passenger Move Charge	$38.00
• U.S. Federal Customs Fee	$13.40
• Canada Goods and Services Tax	$1.10
• September 11 Security Fee	$6.80
• U.S.A Immigration User Fee	$18.80

These taxes are set and apply regardless of the length of the flight. The massive impact of these taxes is really felt on short haul trips. In Australia, Perth based Skywest Airlines had a promotional fare in 2004 on the one hour Perth-Albany route and half the fare of A$99 was taxes. Further perspective can be gained on the impact of taxes from the graph on page 183.

The A380 painted up in Singapore Airlines colors.
Airbus

DETERMINING THE PRICE OF A TICKET

In order to arrive at the price an airline will charge for its tickets, the interplay between two underlying principles are considered.

They are:

- Price elasticity
- Yield management

Price of an airline ticket and the volume of tickets sold are the two basic components that determine an airline's revenue. Finding the optimum balance between the two is essential for an airline to maximize its revenue.

An important factor in determining the price of an airline ticket is what is termed "price elasticity". Price elasticity is a basic principle of economics. It is the relationship between the price of a product and its effect on generating demand for a product or service. Typically, the lower the price of a service or commodity, the more of it will be sold. In other words, for a 10% decrease in price there can be in excess of a 30% increase in demand. However with airline's profitability so marginal, price elasticity is a fine art and requires careful planning.

For airlines, the strategy is to stimulate the market to sell seats at a price that generates demand for a ticket without going out of business. It is absolutely essential that economies of scale are achieved. It can be disastrous to enter a market with low fares against an existing airline's service just to find the route will only support one flight a day. For economies of scale to work, a basic rule-of-thumb is to aim for a service of at least four flights a day, so that the instigation of the necessary costly ground staff and infrastructure can be justified.

The market for air travel can be likened to a pyramid. There are relatively small numbers of people at the top seeking a premium product (first class), larger numbers seeking an appropriate class of travel to their business needs (business class) and various groups that are most likely to purchase cheaper seats (coach or economy class) as a tradeoff for fewer creature comforts.

Today, given the widely differing cost structures among airlines, it is not surprising that a source of irritation to travelers is the myriad fares offered, with big differences between the most and the least expensive. It is not uncommon to discover a flight has 20 fare types.

Yield management is all about the trade-off between selling discounted tickets to fill up a flight and selling tickets at full price. The process takes into account consumer behavior and historical data. Naturally, the time-sensitive business executive will pay more for a flight, while the leisure traveler is driven by other priorities and does not mind traveling at a less popular time. For a higher fare, the traveler gets cancellation options, open return and is able to change their schedule without penalty. The leisure traveler is happy to trade all that for a much lower fare. Airlines also have higher fares for peak season and lower fares to encourage travelers to take to the air in the middle of winter.

As a result of yield management, 90% of the fares sold by US airlines are discounted, with discounts averaging two-thirds of the full fare. An interesting statistic revealed by the ATA is that a relatively small group of travelers - the frequent flyers who take more than 10 trips a year - account for a significant portion of air travel. These flyers represent only 8% of the total number of passengers flying in a given year but they make 40% of the trips.

DECLINE IN AIRFARES

As has been explained in the previous chapters, advances in technology have impacted upon every level of an airline's operation. We have learned about the amazing advances in aircraft manufacturing, the incredible achievements in aircraft efficiency with the progress in engine performance, the decrease in the number of crew required to fly an aircraft, the dwindling maintenance burden involved in aircraft upkeep, the employment of bigger aircraft able to fly longer distances ever more safely, and the implementation of computers to smooth out the effects of variable demand.

All these factors have had a dramatic and continual effect on bringing airfares down. As the airlines have had the benefit of lower operating costs as a result of these technologies, they have been able to pass the cost savings on to their customers.

Politicians, regulatory authorities, the media and the general public have vigorously maintained that airlines are focused on charging the highest possible price for tickets and deliberately ripping-off the public as a result, just to feather their own very profitable nests. This is a stance that is impossible to justify given the number of airline failures and the industry-wide profit figures showing a tiny 1.1% profit margin in the year 2001. What is seldom heard is the truth about how airfares have lagged well behind every other economic parameter - CPI, average weekly earnings and inflation.

The decline in airfares when compared with inflation and average weekly earnings has been nothing short of extraordinary. One example is the North Atlantic route (See Graph 15.2), often considered the world's premium air route and certainly the focus of intense competition and innovation. In 1948, the return fare was about $840 for the first class-only service. In 1952, Pan Am's introduction of tourist class slashed that to $486 at a time when the average weekly earnings were $69 in the US. This meant that the average traveler would need to spend just over seven week's salary to buy a ticket.

In 1959, when the 707 was introduced, the economy class fare was $300 return, while average weekly earnings had risen to $99, representing three week's

Douglas DC-8 Super 63 dominated the North Atlantic routes from the late 1960s till the 747 was established in 1973. The aircraft offered almost 40% increase in payload over the DC-8-50 and 707-320 for only a 5% increase in operating costs.

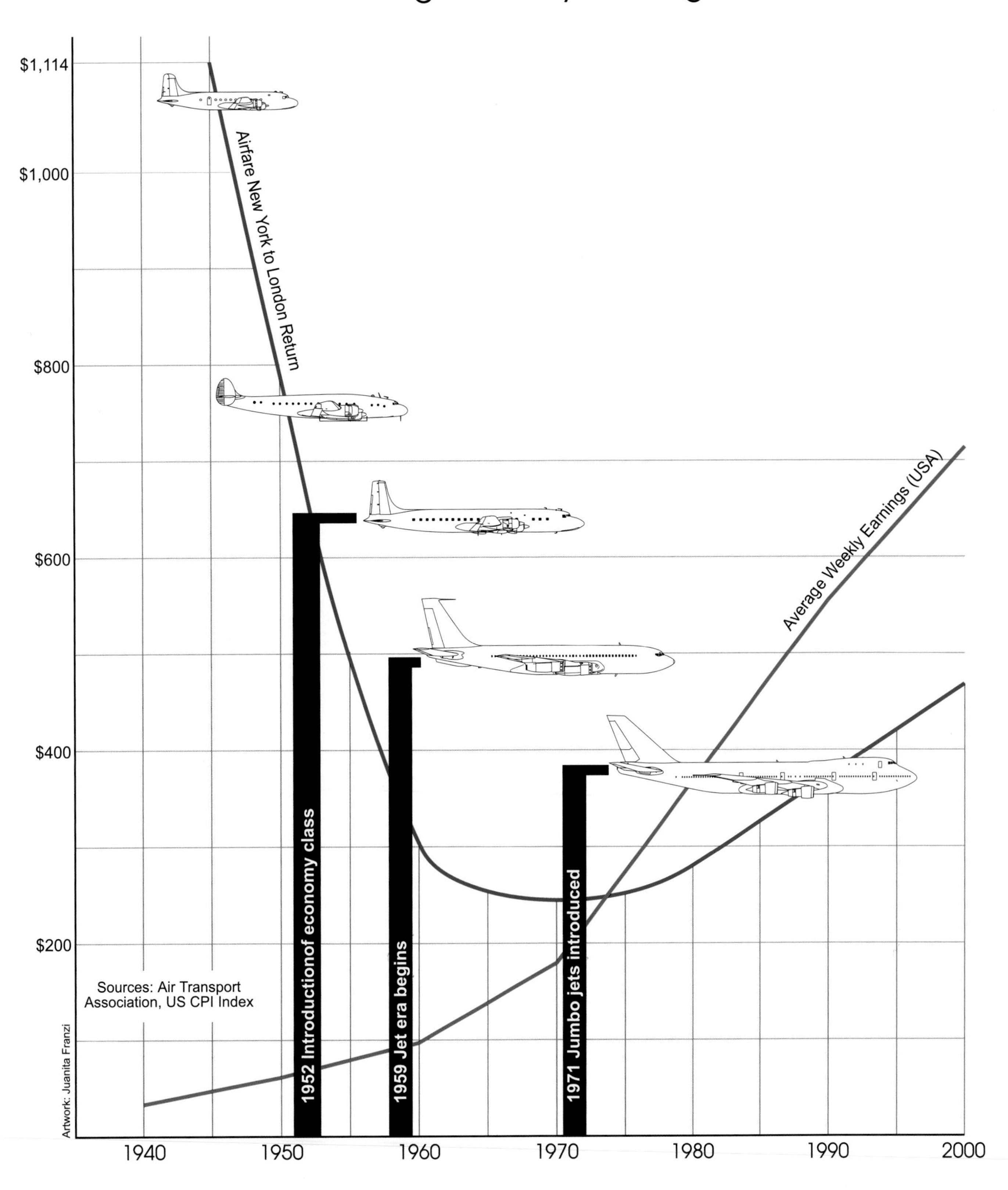
Decline in Airfares
vs
Average Weekly Earnings
$1,114
$1,000
$800
$600
$400
$200
Airfare New York to London Return
Average Weekly Earnings (USA)
1952 Introductionof economy class
1959 Jet era begins
1971 Jumbo jets introduced
Sources: Air Transport Association, US CPI Index
Artwork: Juanita Franzi
1940
1950
1960
1970
1980
1990
2000

Graph 15.2

VC-10s of BOAC plied the North Atlantic routes through the 1960s until replaced by 747s in the 1970s. *BAE SYSTEMS*

salary. The lowest regular fare today is $468 return but, unlike the fare in 1959, this includes $98 in taxes, charges and levies, while the average weekly wage now is $714. Thus, the fare today can be purchased for just over half of one week's salary.

In 1959, the American Transport Association (ATA) in its annual report was able to state that the US Government did not mint a coin small enough to reflect the change in airfares over the previous 10 years. "The average fare for a passenger to fly one mile is now but 3/10ths of one cent higher that it was 10 years ago," the report claimed. As a comparison, during the same period (1949 to 1959), US consumer prices rose 21% and transportation (bus/train) prices rose 61%.

In 1970, the ATA also was able to report that airfares were at that time, the same as they were in 1959 due to increased efficiency of the industry, despite all other costs increasing.

The reduction in airfares, relative to other consumer costs, was illustrated in a snapshot given by the ATA in 1970 when it reported that in 1958 – prior to the introduction of jets – the airfare made up 76% of the cost of a 10-day holiday package, while meals and hotel costs constituted 12% each. By 1970, the balance had swung dramatically to 49% for the airfare while meals and hotel costs accounted for about 25% each. In 2000, the airfare component for a typical 10-day package to Europe from the US was just 20%.

A BOAC Comet 4 was the first jet to cross the North Atlantic in 1959, which started an unprecedented boom in air travel. *BAE SYSTEMS*

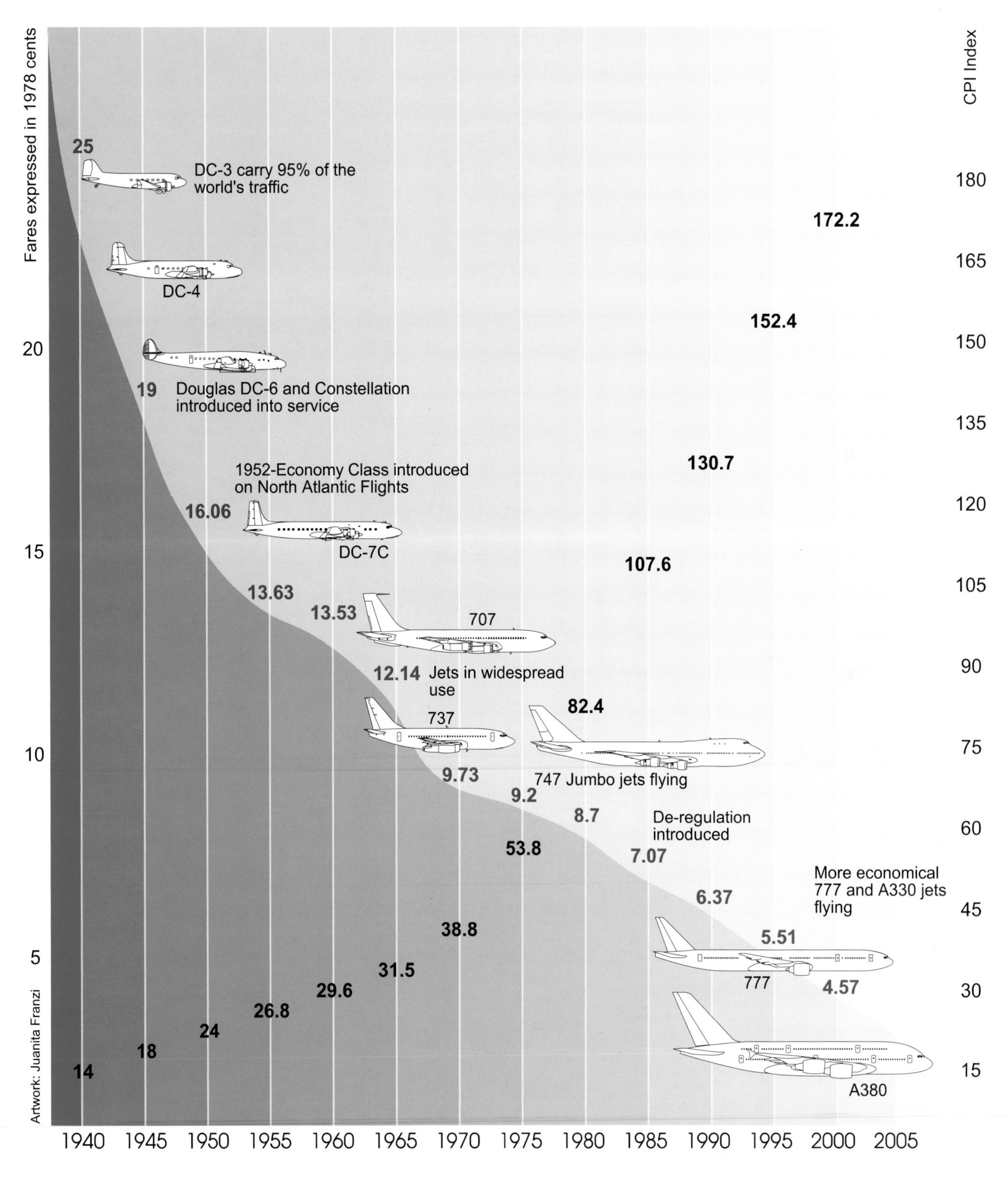
Passenger fares in cents per mile
vs
CPI index
Source: Air Transport Association of the USA
Fares expressed in 1978 cents
CPI Index
25
DC-3 carry 95% of the world's traffic
DC-4
20
19 Douglas DC-6 and Constellation introduced into service
1952-Economy Class introduced on North Atlantic Flights
16.06
DC-7C
15
13.63
13.53
707
12.14 Jets in widespread use
737
82.4
10
9.73
9.2
747 Jumbo jets flying
8.7
De-regulation introduced
7.07
53.8
6.37
More economical 777 and A330 jets flying
5.51
777
4.57
A380
38.8
31.5
29.6
26.8
24
18
14
5
172.2
152.4
130.7
107.6
180
165
150
135
120
105
90
75
60
45
30
15
1940 1945 1950 1955 1960 1965 1970 1975 1980 1985 1990 1995 2000 2005
Artwork: Juanita Franzi

Graph 15.3

500th Douglas DC-6/DC-7 rollout. It was a Pan Am DC-6 that introduced the first tourist fares across the North Atlantic and passengers paid about 15 cents per mile.

YOU CAN'T ARGUE WITH FIGURES

The most compelling picture available on the decline of airfares relative to inflation and in real terms is provided by the US Department of Transportation. We have chosen the US as the example here because the country and its airlines were largely unaffected by damage during WWII and it was the first to introduce deregulation and Open Skies. Similar reductions are found in all countries, although Europe and Asia were ravaged by war.

Indexed figures are available back to 1937 and show that in that year, US airlines charged passengers on average 26.92 cents per mile in 1978 cents for system-wide travel. In 1937, the US Consumer Price Index was 14.4 (1982-84=100). As can be seen from the Graph 15.3 on the previous page, there has been a dramatic reduction in the airfare while the CPI has soared upwards.

Interestingly, in the years following the introduction of deregulation in 1978, there was no change in the trend. In fact, fares actually jumped from 7.81cents per mile in 1979 to 8.70 cents and then 8.85 cents in 1980/81. In the meantime, the CPI had increased to 82.4 in 1980. More than 20 years later in 2001, passengers in the US were paying 13.41 cents a mile while the CPI soared to 177. That trend is repeated across the globe regardless of deregulation or Open Skies.

In the US, the relentless downward trend in the cost of airfares has continued despite shattering events such as the Great Depression and WWII. The same trend can be seen in Australia's cost of living figures in Graph 15.4

And the future fare is the easiest thing to predict. Without question, airfares will continue to decline in relation to average weekly earnings. On the Australia-London route, the downward spiral has been spectacular as can be seen in Graph 15.5. In 1945, the return airfare was the equivalent of 130 week's salary. By 1965 with the introduction of the 707, it had

When the Boeing 747 was rolled out in 1969 the airfares had slumped to just 10 cents a mile.

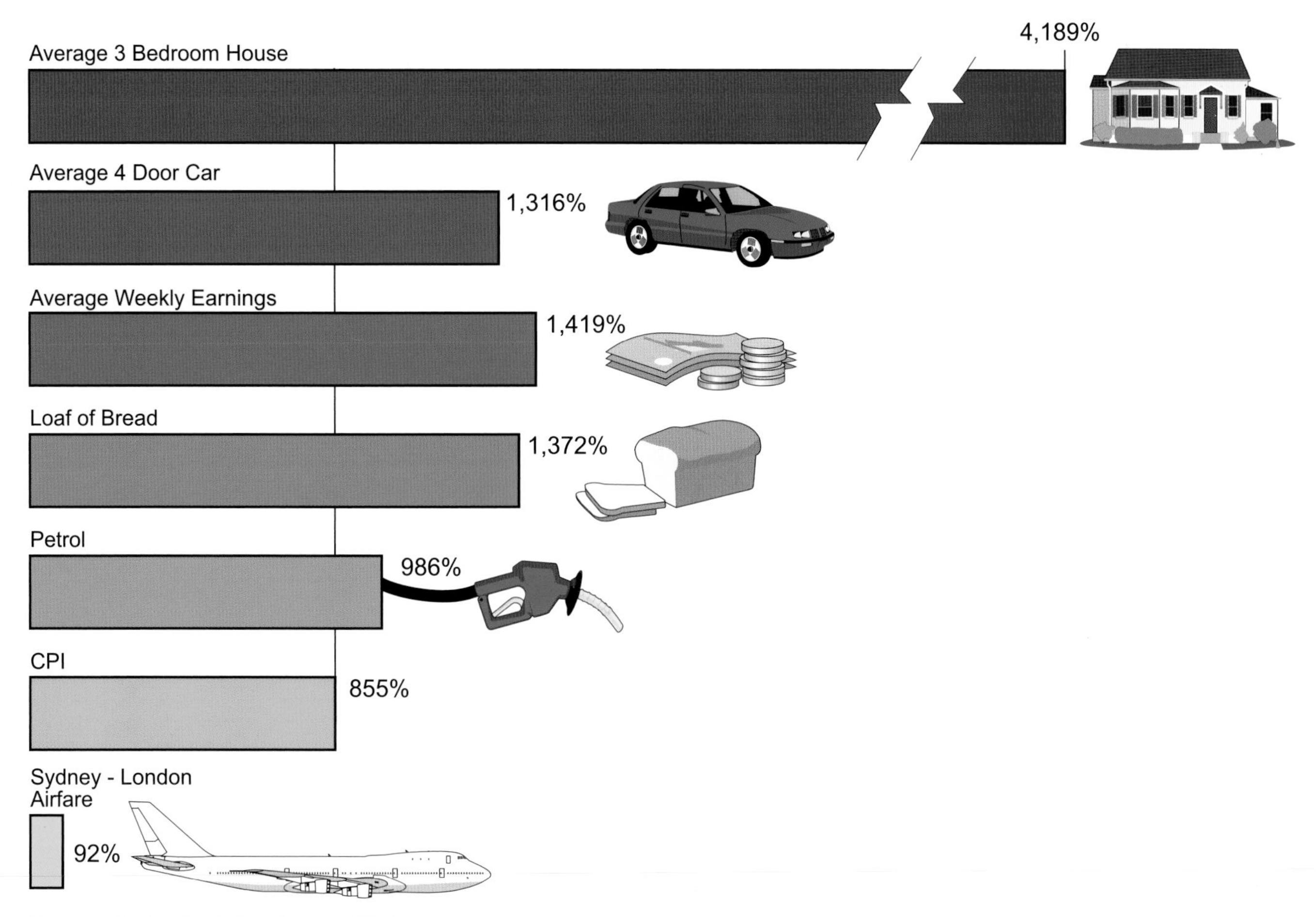

Graph 15.4

dropped to 21 weeks and with the 747 in the early 1970s that figure dropped to eight weeks. In 1991, it was equivalent to five weeks' salary and today it accounts for just over two weeks' average weekly earnings.

In 10 years, analysts suggest that the return airfare from Sydney to London will be just one week's average weekly earnings and half that again 10 years later. Supporting that trend are ICAO figures which show that since 1970 overall global aircraft, fuel, and labor productivity trends (notwithstanding short-term spikes) have all been positive with an annual productivity gain of 5%.

First Qantas Boeing 707-138 lands in Honolulu in 1959.

JetBlue with its fleet of A320s has continued to push fares down.
Airbus

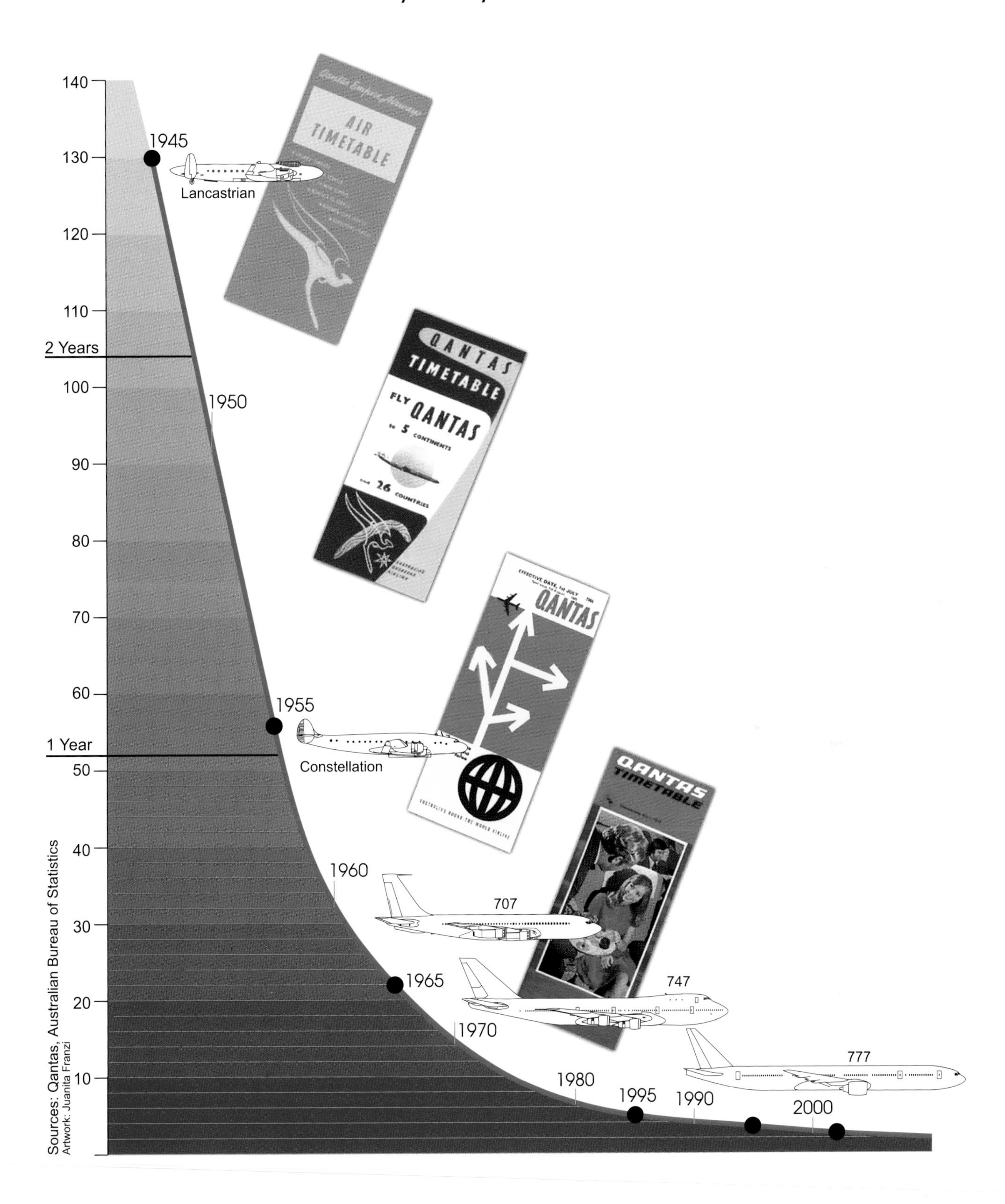

Number of Average Weeks Earnings required to fly from Sydney to London
140
130
120
110
2 Years
100
90
80
70
60
1 Year
50
40
30
20
10
1945
Lancastrian
1950
1955
Constellation
1960
707
1965
747
1970
1980
1995
1990
777
2000
AIR TIMETABLE
QANTAS TIMETABLE
FLY QANTAS
to 5 CONTINENTS
26 COUNTRIES
QANTAS
QANTAS TIMETABLE
Sources: Qantas, Australian Bureau of Statistics
Artwork: Juanita Franzi

Graph 15.5

A factor rarely considered by the traveling public or regulators is that of competition from destinations - particularly for holiday traffic. Often travelers want a certain type of holiday not necessarily a destination in particular, for example, a cruise, a skiing or a tropical holiday.

The Boeing 717 was used to launch JetStar services and this example was adorned with an advert for Budget car rental company. *Misha Popov*

For travelers in Los Angeles a skiing holiday may be at either Reno or Aspen or Whistler in Canada – whichever destination offers the best package. Part of the package includes the airfare. So if an airline has a monopoly to a particular ski resort, by default they must charge a similar fare or lose business to a resort that has four airlines serving it.

In Australia or New Zealand those seeking a tropical getaway, have a host of choices including the Great Barrier Reef in Queensland, Hawaii, Fiji, Tahiti, and Bali to choose from and all destinations have to be competitive in their packaging. For most travelers a beach is a beach as long as it has swaying palms. They will almost always choose the destination which offers the best inclusive deal for airfares plus accommodation plus transfers and/or on-ground tour options.

Many destinations in the Australasian region are only served by one airline because of size but must remain competitive with other major resort destinations such as Fiji and Bali to attract tourists.

The airfare component of a trip has decreased significantly over the past 45 years. That reduction, relative to other consumer costs of a trip, was illustrated in a snapshot given by the ATA in 1970 when it reported that in 1958 – prior to the introduction of jets – the airfare made up 76% of the cost of a 10-day holiday package, while meals and hotel costs constituted 12% each. By 1970, the balance had swung dramatically to 49% for the airfare while meals and hotel costs accounted for about 25% each. In 2000, the airfare component for a typical 10-day package to Europe from the US was just 20%.

With ground costs being such a major component of a holiday package, the overall cost of a vacation is far more dependent on competition between resorts and destinations. As a result, many town councils have removed local airport head taxes to help stimulate traffic. This is the case especially on very short flights where the airport and security taxes can make up to 40% of the airfare.

Another important factor to consider is when two airlines are competing on a regional route where traffic is light, they must each use, for example, a 19-seat aircraft to break-even, whereas one operator with a bigger 36-seat aircraft would be far more economical and be in a position to offer lower fares for their passengers.

What makes up an airfare?

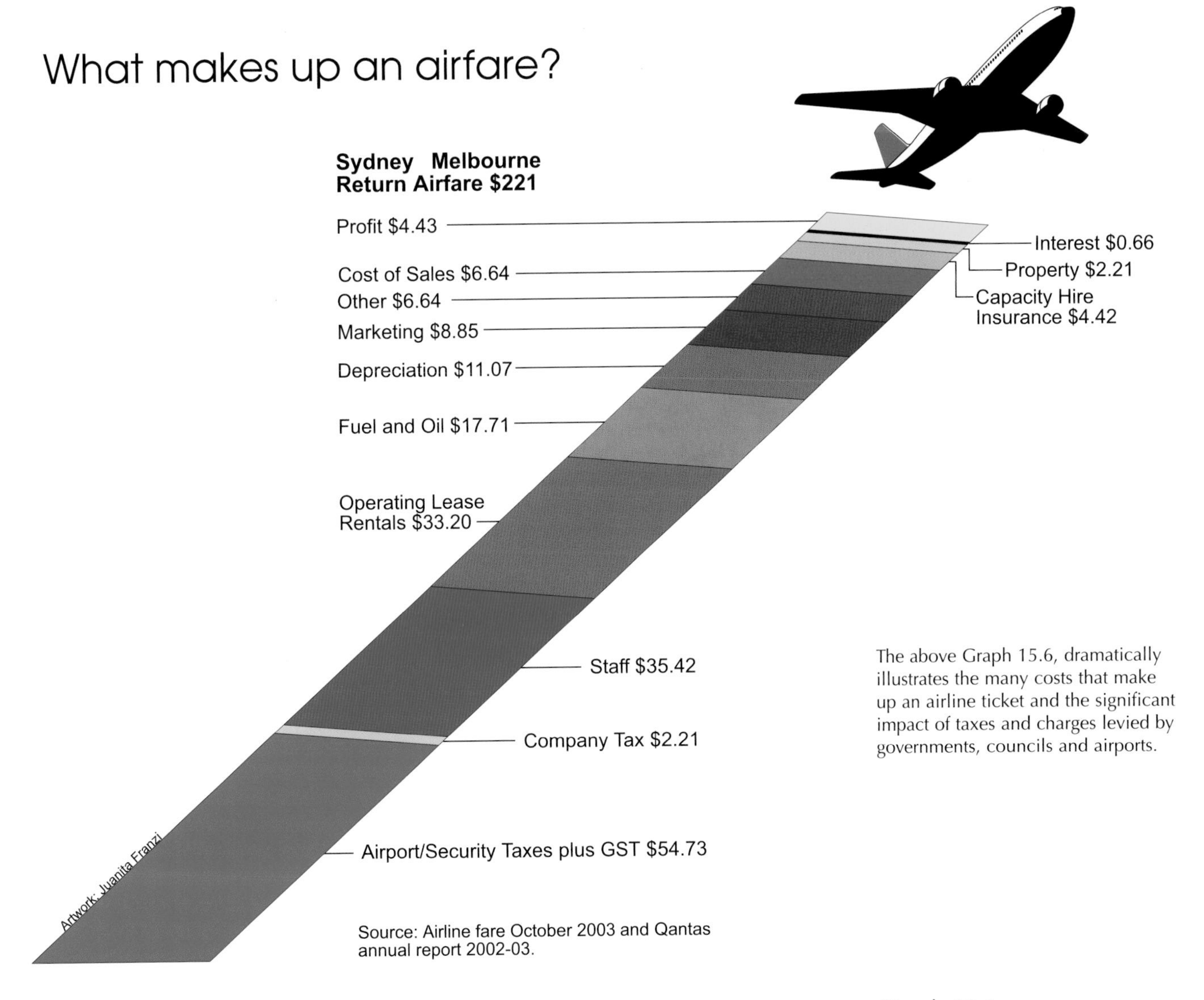

The above Graph 15.6, dramatically illustrates the many costs that make up an airline ticket and the significant impact of taxes and charges levied by governments, councils and airports.

Graph 15.6

The 106-seat A319 bringing holidaymakers into land far more cheaply than two 50-seat aircraft.
Airbus

In Malaysia, new start up AirAsia has forced fares to all time lows. *AirAsia*

The use of internet booking procedures is yet another factor influencing the downward trend in airline tickets which are making air travel increasingly more affordable to a wider range of travelers.

INTERNET FARES

The explosion in the use of the internet has afforded airlines a rapid way to reach the public with special deals while simultaneously cutting out the distribution costs which can account for up to 15% of the cost of the ticket.

Those costs, made up of commissions, ticket offices, etc. are now around 10% and experts suggest that they will drop to 5% in the next five years.

Low-cost airlines, such as Southwest Airlines, have led the way with aggressive use of its website for fare booking. Major airlines have followed suit and have been able to tap into the low-fare market by repositioning their websites to mirror those of so-called low-fare carriers.
Most analysts believe that the role of travel agents for simple point-to-point bookings will continue to decline as airlines offer their cheapest fares on the internet. The role of travel agents is likely to shift solely to coordination of packaged and tailored holidays.

American West, a low-fare carrier in the US, gives a useful insight into the cost of distribution. Its distribution costs of selling a ticket are:

- Traditional travel agency booking $23
- Reservation through the airline's own call centers $13
- A booking made on the website $6

And that is certainly the case in Malaysia where new low-cost upstart AirAsia has started an Internet revolution. According to AirAsia's Chief Executive Tony Fernandes many said the airline "was crazy." But headline-grabbing giveaways and low-fare sales sent Malaysians flocking to cyber cafes to book seats. One grandmother rang her son in London who booked and paid for the seat she wanted. In late 2003 50% of the airline's bookings are made online, an extraordinary feat considering only 18% of Malaysians use the Internet. According to Fernandes every time the airline launches a new promotion, the Internet booking level climbs.

American Airlines 727, which helped to drive fares down in the US, poses with a Ford Trimotor of 1929 vintage. *Boeing Historical Archives*

CHAPTER 16

Alliances

It is obvious we are fighting for the Air France Group. . .
But in actual fact, we are also fighting for France.
– Christian Blanc, Chairman Air France, 1996 –

The KUSS alliance combines to buy commercial aircraft. Here KLM, Swissair and UTA DC-10s are delivered. *UTA*

TO ALIGN OR NOT TO ALIGN

While they are all the rage today and considered by many as a new phenomenon, alliances in one form or another have been around since the 1920s. Certainly, they were nowhere near as sophisticated as today's associations but they did have one very important thing in common and that was their transience.

The transitory nature of alliances has been a fact of life, implicit in every announcement of a new teaming. For instance, in 1997 there were 121 new alliances but 102 were dissolved. Airlines have typically used alliances to further their business interests and develop new routes but as the airline grows or wanes the alliance often becomes excess baggage.

Delta Airlines President and CEO Leo Mullin put another perspective on alliances when he said that they were "an inevitable but not hazard-free" next step in the evolution of the industry. But he said that for Delta, at least, they are "not critical", since the importance of its southeastern US base at Atlanta "overwhelms" anything that happens elsewhere. Other airlines, such as Swissair, before it was restructured, used alliances to sidestep government restrictions that hampered its ability to expand.

Alliances over the years have taken very different forms. The biggest and strongest alliance or cartel was the International Air Transport Association (IATA) which was formed after WWII, largely to protect European airlines against what was seen as the overwhelming dominance of US airlines which were untouched by war. Subsequently, IATA was often criticized for protecting economically inefficient state-owned carriers from the rigors of market competition.

Delta's first DC-8-10 soars in 1959. On alliances Delta Airlines maintains that the importance of its southeastern US base at Atlanta "overwhelms" anything that happens elsewhere.

Through the 1960s and 1970s, the role of IATA changed as airlines around the globe found themselves on more of an equal footing. It was recognized that more liberal aviation policies help promote competition so there has been an unprecedented effort to remove archaic barriers to the expansion of international air services. Today IATA is an aggressive force promoting competition and the advancement of aviation.

Many early alliances were an extension of Colonial control. Yet others were attempts to increase spheres of influence. Some examples are:

- Air New Zealand was formed in 1940 as a three-way venture between the British government through BOAC (British Airways), the Australian government through QANTAS and the New Zealand government.
- Air France helped set up Air Afrique, Royal Air Maroc and Tunisair.
- KLM helped set up Garuda and VIASA. In the early 1960s, one of KLM's DC-8 jets was painted with KLM on one side and VIASA on the other.
- British Airways helped set up Nigeria Airways, East African Airways (Kenya Airlines), Iraqi Airways and Cyprus Airways.
- Pan Am was heavily involved in a number of African and South American airlines.
- SAS assisted Thai International.

Two major technical alliances emerged in the 1960s. The KUSS alliance linked KLM, SAS, Swissair and UTA, and the ATLAS alliance linked Air France, Alitalia, Sabena and Lufthansa to share the maintenance burden for the 747 and DC-10 aircraft. The DC-10 was the first jet to be purchased by a group of airlines participating in an alliance - KUSS and then ATLAS - with only Air France not buying the aircraft. Neither grouping exists today.

Replacing the heavily regulated era under IATA, airlines have sought alliances to insulate themselves from the bigger US operators. In many cases, alliances have been sought with key US airlines such as Northwest, United, American and Delta who are the cornerstones of the current Star, Oneworld, Wings and Skyteam alliances.

An Air New Zealand DC-8-50 rolls out still carrying the TEAL markings, a legacy of Tasman Empire Airways and its joint ownership by the British, Australian and New Zealand governments. *Air NZ*

When Swissair took delivery of the Convair 990A in 1962 alliances usually took the form of helping developing nations establish their national airline. *San Diego Aerospace Museum*

Pan Am was one of the first airlines to help developing countries set up their airlines. Here passengers board a DC-7C in the late 1950s when Pan Am's network circled the globe. *Pan Am*

Today, alliances start off with a code-share arrangement, which allows airlines to sell seats on each other's flights under their own designator code. In the case of connecting flights of two or more code-sharing carriers, the passenger's entire flight is displayed as a single carrier service on the reservation system. From the passengers' perspective, it gives the impression of a service provided by one airline, while in reality several airlines may be involved.

Alliances are, of course, an extremely effective way for airlines to cut overheads when they are not allowed to merge because of national interest barriers. Because of the significant national interest aspect, airline consolidation has been slow. This contrasts with the automotive industry and the oil industry which have worldwide consolidations. The world's six biggest airlines only have 30% of the global market and, by contrast, the petroleum industry's half-a-dozen corporations control nearly 80% of the business. Similarly, in the automobile industry the six biggest players have a 62% share.

The late 1980s and early 1990s saw the growth of new forms of international alliances with very different characteristics to the previous associations in that they served different purposes. As a rule, they have been less institutionalized and formed by mainly privately-owned airlines outside of governmental or inter-governmental agency initiatives. The main growth has been in international alliances, with one of the first of the new breed, between American Airlines and Qantas, signed in 1985. The goals were simple - increase revenue and reduce costs.

One negative impact of code-sharing is that quite often passengers buy a ticket on a particular airline because of its good safety record or because of an aircraft type preference but find themselves on an airline whose safety record is less than desirable or on an aircraft type that they do not like. This may have legal ramifications in the case of an accident.

BENEFITS

Global alliances such as Star, oneworld and Skyteam provide a host of benefits for travelers:

- Code-sharing for seamless travel - where passengers are issued tickets for the whole of their journey from their start point, even though they will be shifting between airlines along the way.
- Linked networks for global coverage.
- Frequent flyer programs - a recent survey found that 94% of respondents belong to a Frequent Flyer program and 60% belong to three or more.
- Airport Lounges - Star Alliance has nearly 500 lounges while oneworld has nearly 300.
- Transferable priority status - if you are a priority passenger with one airline, that priority status transfers to all other airlines within the alliance.
- Joint ground handling, marketing and, in many cases sales offices around the globe.
- Combined ordering of materials and services to cut costs.

Star Alliance airlines are considering an order for a 100-seat regional jet with common fit-out to lower the price. This aspect hopefully will lead to bulk buying of aircraft, which will be a major source of savings for airlines as well as manufacturers.

An interesting example of the extreme cost benefits of buying aircraft in bulk is the United States Air Force's (USAF) purchase of the C-17 cargo transporter. When the USAF was buying in lots of eight each year the cost of each aircraft was a whopping $300 million. However, when it purchased 60 aircraft in a multi-year procurement, the cost dropped to $150 million each.

Airbus and Boeing are able to shave millions of dollars off the cost of buying a jet when they are presented with multi-year orders of 200 or more A320s or 737s respectively. Obviously, airlines working together in alliances to buy a fleet of jets, can gain huge benefits from linking their orders in a single deal for an identical specification with the manufacturer.

Participating airlines in alliances expect to attract more customers because as a group they can fly anywhere in the world and by coordinating schedules and linking frequent flyer programs, the allied airlines create a series of mutually reinforcing incentives for passengers to keep their itinerary within the group. This has become more important as airfares continue to decline and travelers become more adventurous and want more options to visit exotic destinations. The airlines can operate far more efficiently and thereby reduce costs by consolidating flights on over-served routes, combining ticket offices and cross-utilizing sales forces.

The consolidation of flights can take the form of using a 300-seat jet rather than two smaller 150-seat aircraft. In many cases airlines hold cross equity such as British Airways' 17% stake in Qantas, although many of these equity stakes are eventually sold off as the motivation for the alliance loses its gloss because one airline finds a more suitable or attractive partner.

One question that begs an answer is, where are alliances going in the longer term? Every alliance aspires to offer a network of cooperating airlines, which is able to carry passengers from any point in the world to any other point, and back, using only carriers in its own network. So while the alliances are already global, they are striving to be even more so. As such, they are looking for partners that meet three basic conditions:

- They must serve areas that are not already served or are perhaps underserved, by current partners.
- They must not already be partners in a competitive group.
- They must have image and service levels consistent with other partners.

Of course, this is a major challenge for member airlines, some of whom have extremely high standards. As they strive to extend their alliance network coverage to more "corners of the globe", these criteria become more and more difficult to attain.

Lufthansa is a foundation member of Star Alliance. *Airbus*

FARES TUMBLE

While many observers are skeptical of the purported consumer benefits of alliances, a study by University of Illinois economist Jan Brueckner concluded that alliances - especially when combined with code-sharing (as is typical) - do result in significant cost savings for consumers. Study estimates indicate that Star Alliance will save trans-Atlantic travelers $100 million annually in lower fares on multiple-airline itineraries.

Brueckner, an expert on airline economics, also found that alliance carriers in the US and other countries had fares as much as 30% below those offered by non-cooperating airlines. In his report, Brueckner examined more than 50,000 airfares compiled in the fourth quarter of 1999 by the US Department of Transportation's Passenger Origin-Destination Survey and the data included multi-stage trips with thousands of foreign endpoints.

He found that the average round-trip fare was between 8% and 17% lower among airlines that cooperated in a partnership than among those that did not. Further, it was found that fares dropped an additional 13% or more when airlines enjoyed immunity from antitrust laws (competition laws) and were able to charge whatever fares they liked.

This finding is contrary to that put forward by some competition regulators, who insist that only by having a host of airlines competing on a route, can you bring fares down. Airlines, by cooperating, using bigger, more efficient equipment and cutting down on wasteful and often expensive duplication are able to bring significant cost savings to passengers. Brueckner's report found that there is "strong evidence that airline cooperation in the fare-setting process generates substantial benefits for interline (multi-stage) passengers."

MARRIAGE MADE IN HEAVEN?

In 1990, there were 172 alliances of which 82 involved equity of some kind. In 1994, there were 300 alliances between more than 140 airlines and by 2000 that number had grown to more than 500 between 228 airlines.

These numbers support the general concept that airline alliances are vital and that airlines that do not participate may be severely disadvantaged. Many studies have shown that about 70% of the alliances formed since 1990 have survived.

But a number of key airlines, such as Japan Airlines and Emirates, insist that they will not join a major grouping. Instead, they prefer to have a host of associations with a number of airlines from a variety of global alliances. In both cases, these airlines are dominant in their region and believe global alliances can add nothing to their offering.

A significant historic decision, which is likely to affect alliance structures, was reached in June 2003. At this time, the European Transport Council agreed on a package of measures that passes responsibility for conducting key air transport negotiations to the European Commission (EC). In other words, instead of Singapore negotiating with France over rights for its airline to land in Paris, it will be dealing with the EC and the rights would apply to all European countries and vice versa.

Spain's Iberia has joined oneworld. *Airbus*

The EC has also now concluded Open Aviation Area agreement with the US and is planning to initiate negotiations with other countries in order to secure market access for all EC airlines on a non-discriminatory basis. The package of agreements also recognizes the fundamental importance of changing the nationality restrictions that exist in most bilateral agreements. These restrictions have meant that international routes to

A British Airways Boeing 757 lifts off from Aberdeen Airport in Scotland is a very strong crosswind. British Airways is a founding member of oneworld. *Gary Watt*

The sun will probably not set on alliances. *Craig Murray*

and from a Member State can be flown only by air carriers owned and controlled by nationals of that Member State - in most cases, the national flag carrier.

Until now, air carriers could not risk losing their nationality, since they would also lose their traffic rights and thus European airlines have been prevented from taking any restructuring measures, such as mergers or acquisitions that would involve cross-border deals. Under the new general mandate, being broken down, ownership restrictions are. The overall objective of the proposed changes is to enable the creation of a more level playing field for the airlines of Europe and the US. This has allowed cross-border airline mergers in Europe such as the Air France/KLM merger announced in October 2003.

Smaller airlines in Eurpoe have long sought mergers with bigger players, such as British Airways, but have always been thwarted by legal problems relating to sovereign traffic rights. However, the changes in Europe will have significant ramifications for Asian airlines, which now must deal with a block of countries rather than on a state-by-state basis.

Japan Airlines stayed non-aligned for sometime before joining oneworld.

CHAPTER 17

Environmental Factors

You haven't seen a tree until you've seen its shadow from the sky.
— Amelia Earhart —

First flight of the Douglas DC-8-10 in May 1958, with "smoke" pouring from the JT3 engines related mainly to water injection required at take-off. Distilled water mixed with alcohol aided combustion by providing denser air and allowed additional fuel to be injected without exceeding the turbine-inlet temperature limits. But the early DC-8 and 707 used 5,000lbs of water on take-off.

AIRPORT COMMUNITIES DECLARE WAR

There is virtually no airport in the world that does not have a noise problem of some kind, according to International Federation of Airline Pilots Association Principal Vice President-Technical Standards, Paul McCarthy.

This is despite the fact that today's aircraft noise affects only 25% of the surrounding area of 30 years ago. A good example of this is Heathrow Airport where, according to various studies by regulators, flights have increased by 160% since 1975 but the population affected by noise has declined by a massive 90%, thanks to dramatic reductions in engine noise.
Despite perceptions on the ground, the number of people worldwide who are disturbed by aircraft noise has dropped from 19 million in 1970 to just 800,000 today - a 95% decline, according to the FAA.

The Air Transport Association (ATA) estimates that even if nothing further is done to reduce noise, by 2020 there will be a further 40% reduction in the population affected in the US and Europe, as hush-kitted and marginally compliant Stage 3 aircraft are withdrawn from service.

A ban on the operation of what are called Stage 1 jets, such as the noisy Boeing 707 and DC-8 has been in effect since 1 January 1985. In 1989, the US Congress dictated that all Stage 2 jets, such as 727s and DC-9s, were to be phased out by the year 2000 or hush-kitted to comply with Stage 3. Today, the replacement Stage 3 jets include the Airbus A320, the Boeing 757 and the MD-80.

Paradoxically, as aircraft get quieter, airport residents are becoming more vocal. They have swamped authorities with complaints. In Frankfurt during September 2000, there were no fewer than 56,330 complaints.
To deal with such complaints, 400 of the world's 600 major airports, have some form of noise abatement procedures. Two hundred have curfews - double the number in 1990, according to Boeing.

One airport that incorporates many of the most troubling aspects of the noise issue is Sydney Australia's Sir Charles Kingsford-Smith Airport. Noise there has been a big political problem for Federal, State and local authorities because a host of closely contested electorates are located under the flight paths.

The Government has what is termed a Long Term Operating Plan for the airport. That document, thrashed out in 1997, was designed to ensure that aircraft movements are maximized over water and non-residential land, thereby reducing the impact of noise on residential areas.
The key to the plan is a runway rotation system that involves different combinations of runways being used at different times of the day to provide individual areas with periods of "respite" from aircraft noise.

The Lockheed L-1011 Tristar, like other jumbos of the era had a dramatic impact on reducing the noise around airports and communities.

Safety experts and pilots have called upon the Government to abandon or at least modify the complex noise-sharing procedures. Under the direction of the Government, air traffic controllers rotate runways almost without regard for wind conditions. Each runway has a quota of arrivals and departures and controllers are in no doubt what their role is. That is, to direct flights evenly throughout the runway system according to a predetermined roster. The results are even posted on a website.

A Thai International 747-400 touches down at Sydney airport on the cross runway 27. The aircraft has been forced to yaw into the direction of the cross-wind to stay on course. *Craig Murray*

An Airbus advertisement touting the "quite arrival" -the A300

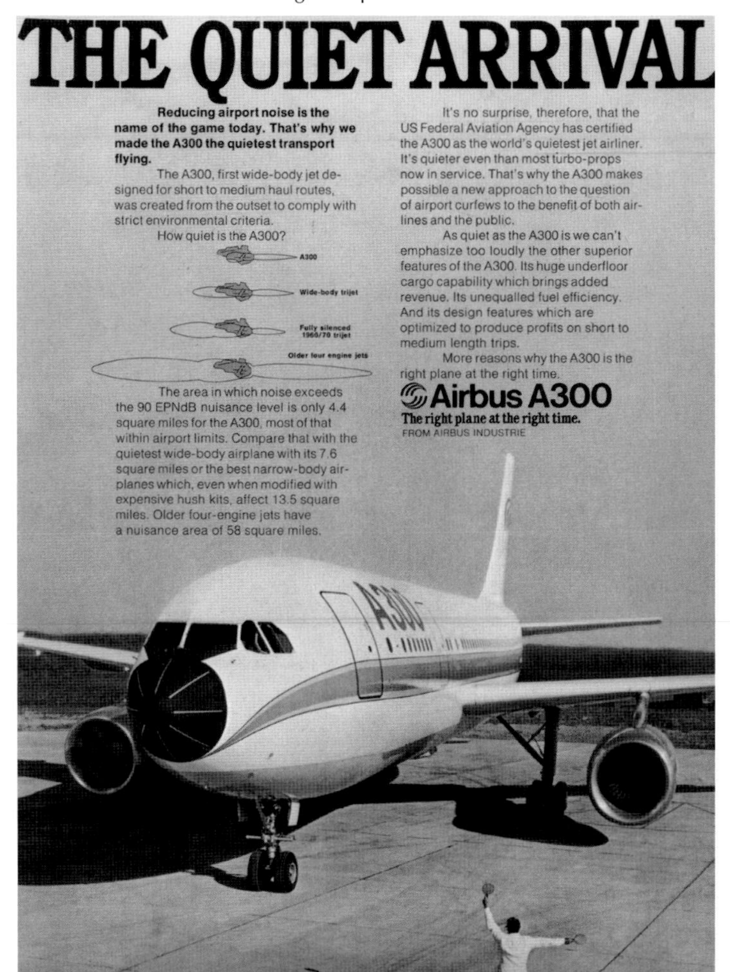

Sydney Airport conducts:

- Intersecting runway operations, where aircraft land and take-off on cross-runways.
- Opposite direction parallel runway operations, where aircraft land from and take-off to the south between 6am and 7am off parallel runways.
- Landings in crosswinds and tailwinds outside International Civil Aviation Organization (ICAO) regulations.

Safety experts claim that noise abatement procedures are a compromise whereby aviation safety is jeopardized for the sake of political expediency. The procedures at many airports around the globe conflict with basic aerodynamic laws, thus degrading the safety performance margin of an aircraft in its most critical phase of flight.

At one stage in 2000, the Australian and International Pilots Association threatened to impose a black ban on the Sydney Olympics to put pressure on authorities to reduce the crosswind component to 20 knots - still 5 knots above the ICAO-recommended maximum. In a strong crosswind, the pilot must point the aircraft (yaw) into the direction of the wind and, in some cases, dip the wing slightly into the wind to stay on course for the runway. The stronger the wind, the more the yaw. Just before touchdown, the pilot straightens

Airbus A300 on its maiden flight
Airbus

the aircraft. If the pilot is landing on a short and wet runway, the difficulty of the landing is compounded significantly. If a tailwind component is added, this forces the pilot to land at a faster speed. The ICAO limit is 5kts downwind, 15kts crosswind. Disturbingly, authorities use the excuse of "it's up to the pilot", but pilots are well aware that if they buck the flow it will mean lengthy delays and possibly missed connections for their passengers.

Numerous solutions have been offered to the vexing problem of aircraft noise. Boeing suggests that airports such as Heathrow can reduce the noise "footprint" (the area affected by noise) of a 747 by almost 20% by using what is termed Continuous Descent Approach (CDA). Many airports require that aircraft approach an airport in what is termed a "stepped approach". This requires pilots to significantly increase power periodically in order to hold altitude at various stages during the landing, as well as when turning, if in a holding pattern. The consequence of this is that the level of noise is far greater than if there was a smooth and gradual descent.
The push by authorities to reduce the impact of noise has led to a number of initiatives. For example, German authorities have examined increasing the glide-slope or angle of descent from three to five degrees which results in a significant three decibel (dB) reduction in noise.

In the United Kingdom, British Airways is trialing what are termed high accuracy, continuous descent approaches to London's Heathrow Airport. The project with the National Air Traffic Service aims to reduce both noise and workload for air traffic controllers. The trial, which started in August 2003, involves the airline's 747s and 777s arriving between 0500 and 0600. Key to the plan is the use of precision area navigation (P-RNAV) flight paths with continuous descent. P-RNAV requires a lateral accuracy of just 1.14 miles (1.85km) compared with the normal 5.74 miles (9.2km). This system will enable the aircraft to maintain cruise altitude longer, reduce fuel burn, noise and air traffic controller workload as there will be no stepped approach, which requires the constant flow of instructions.

Another option is the Quiet Climb System (QCS), which is a software add-on to the Flight Management System and auto-throttles in use today. The QCS on the Boeing 737-800, for instance, takes the guesswork away from pilots by automatically reducing power after take-off to meet noise regulations.

Airlines, fearing penalties of up to $500,000 for exceeding noise limits, have resorted in the past to reducing the aircraft's payload in order to lower take-off thrust and thus comply with noise regulations. At the ultra noise-sensitive John Wayne Airport in Southern California, for instance, QCS has reduced the noise footprint by up to 21%. In fact, Boeing claims that 75% of the 747s departing from Heathrow could actually increase their critical payload by 25,000lb by using QCS.

GREAT STRIDES IN NOISE AND FUEL REDUCTION

The noise generated by engines has reduced by 20dB since the early turbofan engines that powered the 707 in 1961. Compared with the engines that power the A340-500 and the 777 today, that translates into a six-fold reduction in perceived noise to the human ear.

The noise levels of early jets were greater than a noisy jackhammer and were quite distressing. Today, the acceptable noise level, which aircraft must meet, is more akin to standing beside a main road or riding in a car.

As can be seen from Graph 17.1 there has been a dramatic reduction in the noise footprint of aircraft such as the 777. These aircraft now affect only about one square mile (2.58 sq km) surrounding an airport compared to the 54.5 square miles (140.6sq km) for the 707 with turbofan engines. Another comparison is the latest model of the 747, the -400, which affects an area 47% less than the first 747s that went into service in 1971.

But noise is just one aspect of the environmental impact of air travel. Despite its high profile, aviation consumes only 3% of the world's fossil fuels while carrying almost 2 billion passengers a year. If one considers the world's transportation sector as a separate entity, aviation accounts for 12% of fossil fuel consumption, compared to road transport which uses 75% and shipping which accounts for 7%.

In the past 30 years, aircraft fuel efficiency per passenger kilometer has improved by more than 50% through enhancements in airframe design, engine technology and rising load factors. According to Rolls-Royce, more than half of this improvement has come from advances in engine technology. What is more, engines today deliver more than 23 times the thrust and twice the thrust-to-weight ratio of the first jet engines as outlined in Chapter 5.

Boeing's 787 is expected to reduce the impact of noise to just the airport confines. *Boeing*

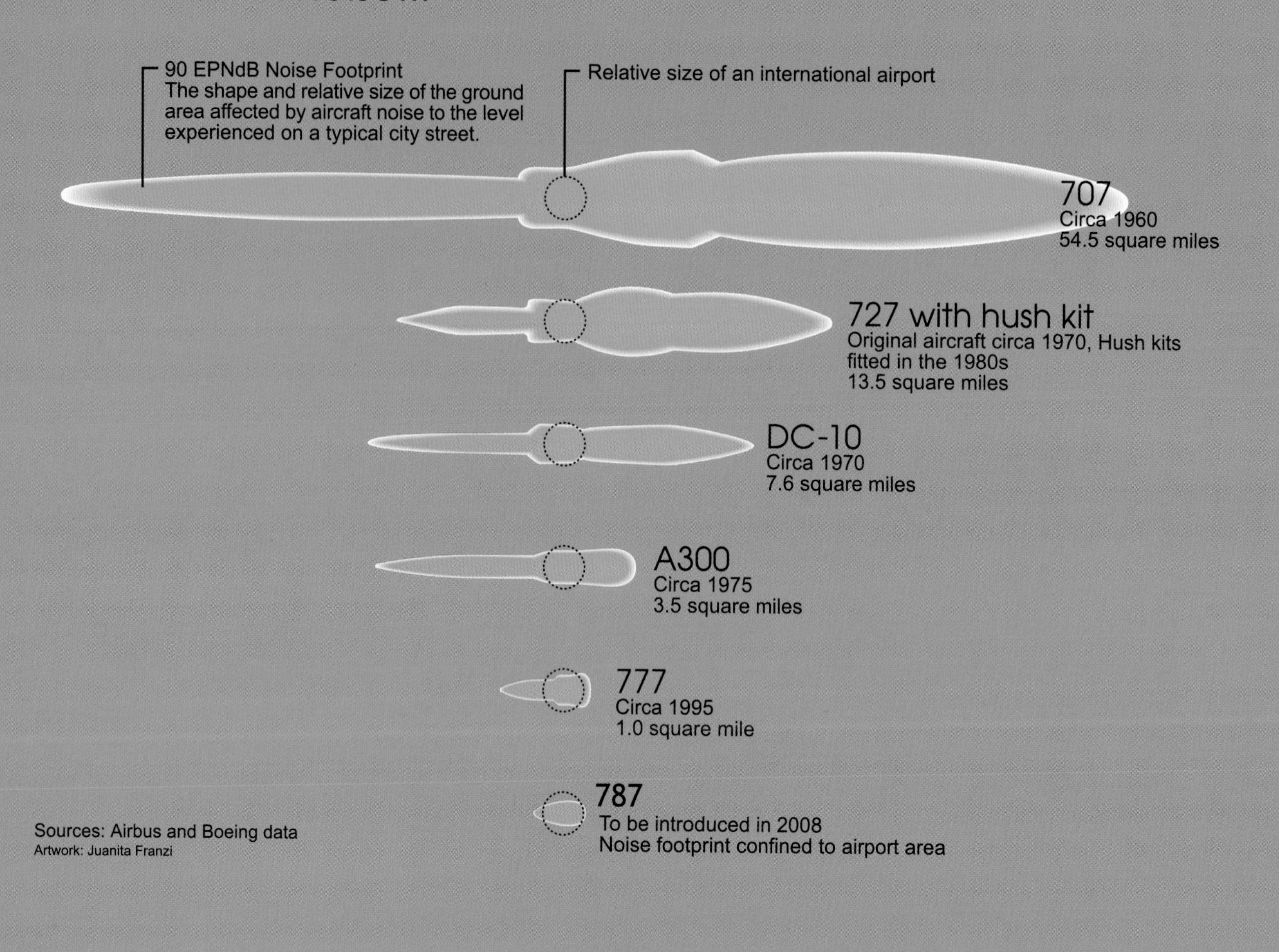

Graph 17.1

Pakistan International Airlines 747-367 in flight over Kazakhstan
Emmanuel Tailliet *"Aviation sans Frontieres"*

ENGINE EMISSIONS AND CLIMATE CHANGE

At cruising altitudes of 30,000-43,000 feet, aircraft produce gases and particles that affect the climate by adding to what is termed the "greenhouse effect", although in reality only a fraction of overall atmospheric pollution is caused by aviation. According to a 1999 report issued by the Intergovernmental Panel on Climate Change, aircraft produce just 3.5% of the man-made emissions that are thought to contribute to climate change.

Since 1970, aircraft emissions per passenger mile have been cut by half. Most emissions are directly related to fuel consumption, and newer aircraft such as the 777 are twice as fuel efficient as aircraft built 30 years ago. Compared with 1950, the reduction is even more dramatic. In the past 50 years, aircraft fuel used per passenger mile has diminished by 70%, according to Boeing.

Many people might think that an aircraft is a gas guzzler compared to a car, while in fact the contrary is true and only when a car is carrying four people does its fuel consumption match that of a 777 and A340 with 75% of seats occupied.

In environmental terms, the emissions from aircraft that are of concern are carbon dioxide, water vapor (in the form of condensation trails) and

Royal Jordanian A340-211 flying New York to Amman just on sunrise.
Emmanuel Tailliet *"Aviation sans Frontieres"*

Unknown aircraft over the North Atlantic takes a turn possibly to avoid some weather ahead.
Emmanuel Tailliet *"Aviation sans Frontieres"*

Air France 747-200 at 39,000ft
Emmanuel Tailliet *"Aviation sans Frontieres"*

A Qantas 747-400 lifts off from the short cross runway 06 at Sydney airport. *Craig Murray*

nitrogen oxides. Engine emissions of carbon dioxide, unburned hydrocarbons and smoke are but a fraction of what they were in 1960. On a global scale, carbon dioxide emissions from aircraft in 1992 were just 2% of the world's man-made carbon dioxide emissions.

In the longer term, various reports have claimed that, despite the massive growth of the airline industry, which doubles every 10 years, jet engines will account for only 3% of global carbon dioxide by 2050.

Limits are in place for emissions from aircraft and the industry has a massive economic incentive to reduce these emissions.

The past 50 years has seen astounding reductions in the fuel burnt per passenger per mile flown and the continual competition between engine and aircraft manufacturers for business will ensure that the trend continues.

According to leading engine makers, a further reduction of 50% in the fuel consumed is possible. Interestingly, this saving is not necessarily all related to the engine.

Boeing's 250-seat 787 will have a 20% fuel saving over competing aircraft made up of:

- 8% from the engines alone
- 2% from the engine/wing combination which is optimized for the new engine
- 3% from reduced weight
- 3% from the advanced aerodynamics and flaps
- 4% from weight reduction from new lighter and simpler systems

"American 51 heavy". An American Airlines DC-10-30 high over the North Pole on-route Dallas-Paris in June 1985.

CHAPTER 18

Sky-high Comforts

The Wright Brothers created the single greatest cultural force since the invention of writing. The airplane became the first World Wide Web, bringing people, languages, ideas, and values together.

– Bill Gates, CEO, Microsoft Corporation –

The DC-2 ushered in a new world of comfort for passengers but reading was the only way to pass the time. *TWA*

Seating onboard a BOAC Lancastrian. *Qantas*

The first hostesses were recruited in the 1930s. *United*

Passengers could pass the time watching the world go by on a Qantas flying boat. *Qantas*

LET'S MAKE TIME FLY

Aircraft today are at the leading edge of technology - the only problem is that we humans are not. Our bodies actually don't like flying in an aluminum tube 30,000ft (9,144m) up at 600m/hr (966km/h) for 14 hours. So to keep us entertained and distracted from our "plight", an entire industry has evolved dubbed In-Flight Entertainment (IFE) which has transformed aircraft into offices and theatres.

Many of us may think that in-flight movies are a recent innovation but the drive to entertain passengers has its roots way back in the 1920s. The 25 April 1925 issue of "Flight International" carried the headline "An Aerial Picture Theatre" and went on to describe the showing of a 1924 film version of Sir Arthur Conan-Doyle's famous "The Lost World".

Across the other side of the Atlantic in 1929, Transcontinental Air Transport, forerunner of Trans World Airlines (TWA), showed cartoons and newsreels. But both events were one-offs and for many years in-flight entertainment remained a hit and miss affair of broadcasting radio.

Passengers also passed many hours in the dining room of the Boeing 314 clipper.
Boeing Historical Archives

The "ultimate bed" on a BOAC Boeing 377 stratocruiser
British Airways Museum

Night cap on a Qantas constellation
Qantas

On the airships of the 1930s, passengers would gather around a grand piano - albeit aluminum to save weight - after dinner. Another media event was in-flight TV on a Western Air Express Fokker F-10 in 1932. However, one of the major problems in the early days was extreme engine noise, which was not a problem in the silent movie era but once sound came to the movies, it became a major hurdle. For the airlines, keeping passengers entertained was a big challenge and quite impossible on the nine-day trip from Sydney to London in the late 1930s.

Hudson Fysh, co-founder of Australia's Qantas, describes travel on the great flying boats: "Getting up out of his chair, a passenger could walk about and, if he had been seated in the main cabin, could stroll along to the smoking cabin for a smoke, stopping on the way at the promenade deck with its high handrail and windows at eye level to gaze at the world of cloud and sky outside, and the countryside or sea slipping away below at a steady 150mph (240km/h) if there was no wind. On the promenade deck, there was also useable space for quoits or even golf, and children could play. There was even a demand for fishing lines at re-fuelling stops, where both passengers and crew members could enjoy the relaxation of dropping a line over the side."

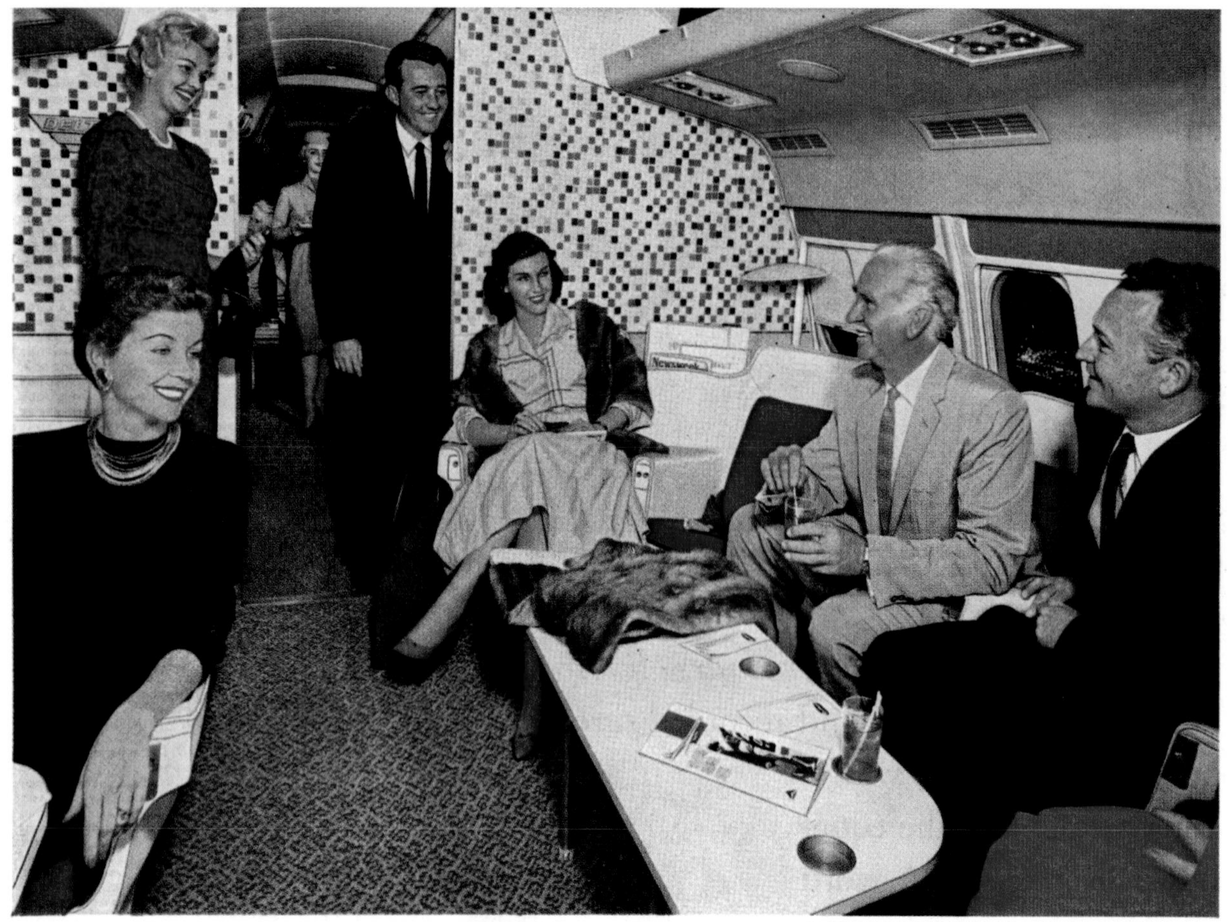

CLUB COMPARTMENT OF DELTA'S CONVAIR 880

The fastest, quietest, most luxurious jetliner travel in all the world is now available as new Delta Convair 880's link New York non-stop with Atlanta, New Orleans and Houston...Atlanta with Dallas...*Chicago with Atlanta, Houston, Memphis, New Orleans. Eighteen major cities will soon be on Delta's Convair 880 and DC-8 jet routes.

*STARTS JULY 1

Advertisement for the Convair 880 in 1961 which painted a picture of opulence and speed.

THE SILVER SCREEN

In 1961, In-flight Motion Pictures developed a 16mm film system for commercial aircraft, which was introduced for first class passengers on TWA but the flight engineers actually spent more time fixing it than monitoring engine instruments. The engineers also had to change the reels.

Aircraft manufacturers have suggested various ways to introduce lounges under the main cabin. Here is a Boeing mock-up for the 747.

This Tristar underfloor lounge was a feature of San-Diego based Pacific South-West Airlines and German charter operator LTU.

The film to be shown on a TWA 707 was MGM's "By Love Possessed" starring Lana Turner and Efrem Zimbalist Jr.
But not everyone was happy with the films, with some passengers complaining that they had no choice but to watch the film, while others complained that they were forced to pull down their window shades. There was also resentment from economy-class passengers who had nothing to watch. TWA responded to this problem by showing movies to all passengers but charging for headsets in economy.

TWA also ran into strife with overseas airlines, which complained that it was unfair for the airline to show free movies when they didn't have movies at all on their flights. There were some radical suggestions that TWA would have its overseas landing rights suspended by some countries unless they charged passengers.

American Airlines was the first to mount a counterattack with a Sony system, dubbed Astrovision, that featured small TVs located between seats so that first class passengers could look down at the screen. Coach passengers had a TV monitor mounted every nine rows through the cabin. These were the first "seat back videos" and American even had nose-mounted cameras to give passengers a view of the take-off and landing.

In a 1964 advertisement, it even promised passengers that its systems would not require the window shades to be pulled, although that aspect never seemed to be carried through.

There were, however, a host of problems in those early days and the systems gave way to the wide-screen opportunities offered by the 747 in 1970. What to do with the space opportunities on the 747 also spawned a rash of suggestions that included McDonald's fast food outlets, hair salons and downstairs lounges. Boeing even built a mock-up of the lounge but there were no takers. However, American Airlines did install a piano bar in the rear of its 747s and DC-10s, but it didn't last long.

In 1971, Trans Com developed an 8mm film cassette that could be changed by flight attendants and in 1975 Texas-based Braniff Airways introduced Atari video games on flights.

On the 747 the upper deck was initially used as a lounge but after a few years that gave way to revenue seats. *Qantas*

Bell & Howell's Avicom Division pioneered video systems in 1978 and a year later electronic headphones were introduced by Pan Am, Air Canada and Air France.

Not quite in-flight entertainment but essential to many, telephones made their debut in 1984 on American Airlines.

Developments now came thick and fast with audio player systems (1985), in-seat video with a 2.7-inch display (1988) and noise canceling headphones (1989).

American Airlines introduced fleet-wide on its 767s, seat-back videos for first class and in 1991 the people's hero, Sir Richard Branson, introduced seat-back videos for all passengers. Singapore Airlines added in-flight fax machines two years later, which were not exactly entertainment but may have given some business executives peace of mind to enjoy the in-flight video.

McDonnell Douglas proposed downstairs seating for their MD-11 but there were no takers.

The business executives really appreciated in-seat power for PCs, which was introduced by Delta Airlines in 1996 and in the same year Delta showed live TV from the Atlanta Olympics.
Swissair upped the ante the following year, installing an interactive Video-On-Demand (VOD) entertainment system on its MD-11s. And going even further JetBlue, the New York-based, low-fare airline, introduced live (via satellite) in-flight TV fleet-wide in all cabins of its A320 fleet in 1999.

DVD players made their appearance in 2000 on American Airlines and Swissair, and email transmissions were demonstrated by Air Canada during a press event on 17 January of the same year. Today's passengers on leading airlines are surrounded by technology and, after some significant teething problems most systems now demonstrate excellent reliability.

The most comfortable economy class seat in the world is on

THE B·O·A·C VC10

The BOAC VC 10 brings a new luxury to Economy class travel, with a completely new type of seating, designed and built by Aircraft Furnishing Ltd. Seating that assures you a much more comfortable flight. You get extra legroom—better support—deeper cushioning—improved sleeping position—personal tables. Just sit back and enjoy the wonderful BOAC cabin service. Pay the lowest fare and fly in greater comfort than ever before. Economy Class on the VC 10!

ALL OVER THE WORLD

B·O·A·C

TAKES GOOD CARE OF YOU

BRITISH OVERSEAS AIRWAYS CORPORATION

Triumphantly swift

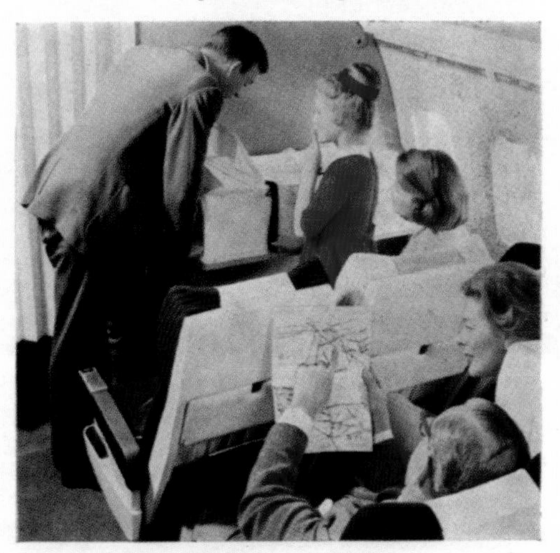

silent

serene

British Airways VC 10 advertisement in 1964, extolling the virtues of flying on the VC 10.

Singapore Airlines new business class – launched in 2006 – sports the widest seats in the industry.

According to the World Airline Entertainment Association, the following systems are available and in service with airlines.

Personal "Distributed" Video: At-seat individual monitors that offer 6 to 24 video channels and up to 72 audio channels of programming, distributed on a pre-established cycle to all passengers from a central system.

Audio/Video-On-Demand (AVOD): A wide selection of digitally stored audio and video content that passengers may independently "order up" at any time (as opposed to "Distributed Video", which provides a pre-established cycle/schedule of programs) from their in-seat monitor. Passengers may also stop, start, pause, rewind programming and access other interactive features.

In-flight Satellite TV: Real-time and/or live audio/video programming fed directly to the aircraft via satellite where passengers may independently access from 25 to more than 100 channels of digital-quality audio/video entertainment (similar to, or identical to, the programming they receive via home satellite dishes).

In-flight Intranet/Email Access: Using the passenger's notebook-computer or an installed interface, the passenger may access up to one million web pages stored on a cabin server (which is updated just before the flight and/or periodically during the flight). Passengers may also send/receive email (which is forwarded/received periodically via an air-to-ground satellite link to the Worldwide Web (WWW)).

Real-Time In-flight Internet/Email Access: Using passenger's notebook computer or an installed interface, the passenger may directly access the WWW and send/receive email, all in real-time, via an air-to-ground satellite link to the WWW.

First class on the A380
Airbus

Seat-back videos on a Cathay Pacific A340
Cathay Pacific

Passenger accesses internet using the Tenzing/Airbus system. Note mood lighting.
Airbus

ASIA LEADS

Ever since Malaysia-Singapore Airlines (MSA) snubbed its nose at IATA in the late 1960s and offered free drinks to economy-class passengers, Asian airlines have been at the forefront of in-flight innovation. Typically, this has been driven by the fact that most of the region's airlines operate a number of long-range sectors to the world's commerce centers of London and New York.

Qantas introduced business class in 1979, while Singapore Airlines and Cathay Pacific have been trendsetters with not only a high level of in-flight meal service but high-tech gadgetry. This is not surprising given that almost half the revenue of some of these airlines comes from the front end of the aircraft and the battle for just 1% increase in business is intense.

So important is it that Australia's Qantas, despite all the post-September 11 problems, has just

The new Emirates First Class *Emirates*

Staying in touch. Seat back video and console. *Emirates*

launched yet another upgrade of its business class with what it terms the "best bed in the sky" that will rival most airlines' first-class offerings. Qantas CEO Geoff Dixon, explains that while first class may be waning, the airline will not remove it until his archrivals such as Singapore Airlines do.

But the cost of staying ahead of rivals is astronomical. Some estimates put the spending for the industry on IFE by 2005 as high as $4.37 billion. To add perspective to this, before 1994 only 40% of 747s were fitted with in-seat videos, whereas today the number is above 95%. But there is a savage weight penalty for the systems. On an Airbus A340, a full state-of-the-art system adds 2.5 tons and on a 747 that figure can be as high as 4.5 tons. That is the equivalent of 24 fare-paying passengers.

Airlines leapfrog each other with the latest innovations in passenger comforts. While Asian airlines have been extremely aggressive in the field, it was British Airways that first introduced first class and business class beds. The airline was also one of the first to introduce a premium economy product for its long-haul fleet.

Singapore Airlines, Cathay Pacific, Qantas, Emirates, Virgin Atlantic, Japan Airlines and ANA are all at the forefront of introducing innovations throughout the aircraft cabin. Typically all passengers have seat-back entertainment systems, while premium class passengers have in-seat power in seats/beds.

Singapore Airlines introduced its A340-500 in early 2004 with just business class and executive economy class which has one seat less across the width of the cabin and an additional 5 inches (12.7cm) of legroom. This new product reflects the increased endurance of the aircraft - up to 18-hours non-stop. This trend to offer specialized aircraft and interiors to suit varying markets is expected to develop further.

CHAPTER 19

Future Flight

Our job is to keep everlasting at research and experimentation, to adapt our laboratories to production as soon as possible, and to let no new improvement in flying and flying equipment pass us by.
– William E. Boeing, founder The Boeing Company, 1929 –

Airbus A380 sporting Qantas' titles for the celebrations at Brisbane, Australia marking the airline's 85th birthday in November 2005.

IMAGINATION

The imagination of futurists seems boundless and aviation is fertile ground. Over the years, we have seen everything from giant flying wings with nine decks for 451 passengers and 155 crew to hypersonic passenger planes that would fly from Sydney to London in just two hours.

The most ambitious was without doubt Norman Bel Geddes' Flying Wing or Flying Sip of 1929. The massive aircraft was to weigh 700 tons, be powered by 20 engines and fly at just 90mph (145km/h). There was to be a 200-seat restaurant, gym, games deck, cafe, bar, solarium and, rather than a cockpit, a navigation bridge. Needless to say, the aircraft never left the drawing board.

Not quite as fanciful, but still very impressive, are the special, high-density 747 aircraft of today that carry almost 560 passengers for Japan Airlines and All Nippon Airways on domestic routes, and Airbus' bold construction of the world's biggest passenger jet, the A380, which is expected to take mass travel to a new level.

The aircraft comes together with components from dozens of locations around Europe and the world. The first aircraft was rolled out in January 2005.

The giant jet is awesome. There are one-piece skin panels that are 116ft long and a flap-track fairing that houses the system supporting the landing flaps that would make a good-sized yacht. Its colossal size is going to translate into more comfort for passengers, claims Airbus.

A380 Cross-section

A380 in Emirates colors at the Dubai Air Show in November 2005.

Airbus also claims the A380 will be about 15% cheaper to operate than a 747. Key to the A380 is its giant wing measuring 9,000sq ft, (836sqm) which is almost double that of a 747 at 5,600sq ft (520sqm). The rationale for the giant wing is to ensure that the A380 can be stretched to carry more passengers, plus it is instrumental in meeting the requirement of reaching a cruising altitude of 35,000ft (10,668m) within just 229 miles (370km) or 30 minutes from take-off.

According to Airbus, a massive wing has some advantages and then some drawbacks. The A380 will land at a gentle 143 knots and will use less runway than a 747. The simple one-piece flaps cut down on what is termed airframe noise and the bigger wing means less thrust for take-off and climb, reducing the noise impact on local communities.

However, the downside is that the wing is heavy and big on drag though not nearly to the extent as first thought. Airbus designers have used new computational fluid dynamics software to achieve a better-optimized wing design than was possible in the past.

To meet the noise restrictions of London's Heathrow Airport, engine makers have increased the engine fan size and thus the bypass ratio of the engines, from the 5:1-6:1 of most big turbofan engines to an 8:1 ratio.

While the A380 is huge and will weigh 606,000 lb empty, Airbus has been instrumental in generating a host of weight-saving initiatives. Some of these weight savings are:

Airbus A380 poses with a Lockheed Constellation at Qantas' maintenance base in Brisbane Australia in November 2005. The two aircraft were in Brisbane to celebrate Qantas' 85th anniversary.

A380 over Touloose
Airbus

- The A380's center wing box, the structural "heart" of the aircraft, is constructed from relatively lightweight carbon-fiber composite material, saving 4,000lb (1,818kg).

- Upper fuselage skins are the first large-scale application of a new class of composite-metal hybrid material called GLARE, developed by the Netherlands' Delft University of Technology. GLARE comprises a layer of fiberglass between two thin sheets of aluminum alloy, which not only saves weight but is highly resistant to the development of cracks. As well as being about 10% less dense than aluminum - for a weight-saving of about 1,760lb (800kg) - GLARE has proven to be superior in terms of fatigue as well as fire and damage resistance.

- Another weight saving feature is the use of increased pressure for the A380's hydraulic systems. For the first time in civil aviation, the A380's hydraulic systems will use increased pressure of 5,000 pounds per square inch (psi), as opposed to the traditional 3,000 psi. This increase in pressure allows the necessary pressure to be transmitted with smaller piping and hydraulic components. The reduction in the size of components, unions and piping not only lowers the weight of the aircraft by around one ton but also improves its maintainability.

Three A380's taxi out at Toulouse.
Airbus

As long as we have looked upwards to the sky and longed to fly, we have also longed to fly faster. But in commercial aviation, flying supersonically or near supersonically has met financial disaster. Billions of dollars have been poured into research into a supersonic transport to replace the Concorde but a host of environmental and economic problems have grounded the dreams.

Boeing's Sonic Cruiser was grounded by the demand for no-frills and low-costs. *Boeing*

The most recent attempt was Boeing's Sonic Cruiser. Not since the 747 blitzed the market in the early 1970s had an aircraft promised to dominate the market. But many asked whether Boeing could make it work. Skeptics, and there were plenty, wondered if the Sonic Cruiser was just a tactic to disrupt Airbus' sales efforts with the 555-seat A380.

Boeing was attempting to build an aircraft that would cruise comfortably where most aircraft don't want to go - the edge of the sound barrier - and weigh as much as a 777 but haul a 767 load without a fare premium. This was a very tall order but conservative Sonic Cruiser Program Vice President and General Manager Walt Gillette was adamant that Boeing had solved all the problems.

However, after the events of September 11, the entire market focus shifted away from high-flying, high-profile travel to budget low-profile airlines. Boeing found that as it referenced all the new technology and systems for the Sonic Cruiser to a conventional baseline aircraft, airline executives started to focus on cost savings rather than speed. Boeing shelved the Sonic Cruiser in 2003, and with it, any chance of supersonic travel or near supersonic travel for at least the next 20 years.

But deep within Boeing's Product Development inner sanctum, another top secret project ran alongside the Sonic Cruiser (Project Glacier). Following the same naming protocol as Glacier, the initiative was named after another US National Park - in this case Yellowstone.

It was revealed in May 2001 but had already been underway for at least four years. The basic plan involved applying as many of the advanced manufacturing, design and systems technologies that were being studied for the higher speed airliner to a conventional design. The 787 was born.

Air New Zealand is a launch customer for the 787.
Boeing

787 - SON OF THE SONIC CRUISER

When Boeing launched the 787, at the same time as it shelved the Sonic Cruiser in early 2003, there were many who wondered if the plane-maker had lost the "right stuff". But British Airways Chief Executive Rod Eddington when referring to the shelving of the Sonic Cruiser, summed up the view shared by many airline leaders by saying, "The aviation lover in me wishes it were the other way around but the airline business person in me believes it is absolutely the right call".

While the 787 lacks the air show excitement of a Sonic Cruiser, it has the potential to dominate the market with its range/speed/economy package. Boeing plans to launch the 787 in the first quarter of 2004 and have the aircraft in service by 2008.

The 787 will come in three models and the base version will carry 200 passengers in a three-class international layout, while the stretched model will carry 250. The 787 will be able to fly a distance of 9,565 miles (15,400km). At that range, the 787 could fly from London to any point on the globe except the South-East corner of Australia, and New Zealand. The short range version, modeled on the base aircraft will carry 300 passengers 4,000 miles (6,500km).

For passengers, the 787 will be the first aircraft of that size to offer virtually non-stop capability around the globe, meaning greater flexibility for airlines to introduce more non-stop, city-to-city routes. That non-stop capability, combined with the aircraft's 20% lower costs, means significantly reduced airfares.

The cross-section of the aircraft will be an eight-across economy configuration. But rather than a circular cross-section, the 787 will sport a double-bubble fuselage with a crease line at the floor. Below floor level the cross-section will be virtually identical to the A330 to take two LD3 containers but above, it will widen to allow for seating and aisles slightly wider than the 777.

Operating Empty Weight (OEW) of the 787 will be 22,000lbs (10,000kg) lighter than its competitor the A330-200 and its wing will be 10% smaller. Key to the performance, as always, is the engines which will have a bypass ratio of 12 - double existing engines - and will be rated at about 70,000lb of thrust. Speed will be similar to the 777.

Passengers will arrive in better shape on the 787 as Boeing plans to decrease the degree of cabin pressurization by about 25% and increase cabin humidity

The passenger entry area of the 787 sports a skylight styled lighting treatment. *Boeing*

from 5% to 30%, which will almost certainly mean an end humidity levels can be increased because the 787 will be built mainly of high-strength, non-corrosive composite material. In conventional aircraft, built of aluminum and titanium, water vapor condenses on the side walls of the cabin and flows into the bottom of the fuselage where it can corrode the metal structure. Boeing plans to reduce the amount of aluminum in the 787 to just 15% - down from 70% on the 777.

Co-author Christine Forbes Smith gives perspective to the 787's huge windows.

In addition, engine noise levels will be greatly reduced. In fact, the aircraft may be too quiet, the designers say. Engineers are apprehensive that because the aircraft may be almost "library quiet", passengers will be concerned that their conversations could be overheard.

But residents around airports are going to love the "library quiet" characteristics of the new aircraft because the area affected by noise will be confined to the runway from which the 787 is taking off.

Northwest Airlines ordered up to 60 Boeing 787-800s on 9 May 2005. *Boeing*

The 787 is the second part of Boeing's strategy to reduce the number of aircraft types it offers to the market. At the top end, it believes its 777-200 and -300 series that can carry up to 365 passengers in an international three-class configuration will dominate, while the 787 will corner the 200 to 250 seat market, it says. The 100 to 200 seat market, dominated by the 737 and A320, will be the next area of focus with Boeing and Airbus expected to launch a new series of models using 787 and A380 technology and systems. There will be a higher level of commonality between models, so much so that certain items may be built on common production lines.

First Boeing 787 fuselage test barrel made entirely of carbon fibre, which is lighter and stronger than the traditional aluminum. *Boeing*

Blended Wing Body may be launched first as a military transport before passenger trials are started. *Boeing*

WINGS WITH PASSENGERS

There is little doubt that the most promising of all concepts of the future is what is termed the "flying wing". Rather than a fanciful idea, the flying wing dates back to the 1920s and was first flown as a bomber in 1947.

The Northrop XB-35 was ordered by the US Army Air Force (USAF) in November 1941 and a further contract was awarded in 1942 for 13 YB-35 service test aircraft. A further order for 200 production B-35Bs was placed in June 1943. But it soon became apparent that the aircraft would not be available for war service and the new jet aircraft on the drawing boards made the piston-engined XB-35 look obsolete. The USAF however, continued with the XB-35 program but only for test flights. The XB-35 made its first flight on 25 June 1946 but both XB-35s were scrapped in 1949. The performance of the aircraft was stunning for its day - the XB-35 could carry a 16,000lb (7,270kg) load out to 9,364 miles. (15,077km)

Carrying on the concept, in the mid-1990s McDonnell Douglas (now part of Boeing) launched a study into the flying wing or Blended Wing Body (BWB) as it became known. The BWB concept represents a revolution in subsonic transport efficiency and NASA sponsored an advanced concept study to demonstrate its feasibility in the late 1990s.

The study compared the 800-passenger BWB and a conventional configuration aircraft over an 8,050 mile (12,960km) design range, where both aircraft were based on technology for a 2010 entry into service. Not surprisingly, the BWB showed some revolutionary advances over the conventional aircraft.

Compared to the baseline aircraft, the BWB had:

- 27% lower fuel burn
- 15% lower take-off weight
- 12% lower operating empty weight
- 27% lower total thrust
- 20% higher lift versus drag

And according to Boeing studies, the BWB would have operating costs half those of a similar-sized aircraft.

The BWB would feature two decks with passengers seated across the width in 10-passenger bays. There would be windows in the BWB's wing leading edge and all seats would have videos with extensive outside views. Cargo and passengers' baggage would be carried out-board of the passenger compartments and fuel would be carried beyond the cargo.

The BWB's three engines would be located atop the aircraft and thus the BWB would be exceptionally quiet with the wing area masking the engines' noise signature. Compared to conventional cylindrical tube fuselages, the center-body pressure vessel of the BWB would be far stronger, greatly improving chances of survival in a crash.

But despite the compelling economics of the aircraft, neither Boeing nor Airbus has it in their plans, as deep concerns exist about the acceptability of the design to passengers. It may evolve that the BWB will find its way into the passenger market through the military route as a tanker and as a cargo carrier.

One concern for designers is that because the aircraft is so wide, passengers sitting in the outside seats could feel as though they were riding a roller-coaster as the aircraft banks and turns after take-off or for landing. However, NASA has tested this problem with a motion simulator and concluded that the ride would be quite similar to a trip in a 747.

PELICAN

The grand prize for the biggest aircraft ever devised is without doubt the Pelican Ultra Large Transport Aircraft (ULTRA) which would be twice the size of the Russian An225 and carry five times its payload - up to 3,080,000lb (1,400,00kgs).

Designed primarily as an over-water aircraft, it would fly at just 20ft (6mtrs) above the waves, taking advantage of an aerodynamic phenomenon called "ground effect" that reduces drag and fuel burn. When over land, the aircraft would climb to 20,000ft (6,096m). With a payload of 1.5 million lbs (680,388kg), the Pelican could fly an incredible 11,500 miles (18,515km). The future of the Pelican lies with military applications and the design is being offered to the US Army to solve its mobility shortfall.

CHAPTER 20

Future Growth

I confess that in 1901, I said to my brother Orville that man would not fly for fifty years . . . Ever since, I have distrusted myself and avoided all predictions.
– Wilbur Wright, in a speech to the Aero Club of France, 1908 –

THE SKY IS THE LIMIT AND THE SKY IS LIMITLESS

That air travel will continue to grow at an extraordinary rate there is no doubt. There are however differing views of how the growth patterns will evolve.

This chapter examines the growth trends, the factors that impact that growth and the influences that impact on the way traffic flows are expected to evolve.

737s line up for take-off

Like telephone lines, cable and satellites, airlines have become a critical and ever growing part of the global economy. In 2005, over 2 billion passengers and 40%, by value, of the world's manufactured exports were moved by air, according to ICAO.

A 2000 report on the Economic Benefits of Air Transport reveals that, since the first jetliner flew in 1949, air travel has grown 70-fold. And in 2000, there were over 18,000 aircraft operating into and out of 10,000 airports.

According to Official Airline Guide (OAG) figures, the growth of cities connected by air, dubbed city pairs, is relentless. For routes with jets over 100 seats, the number of city pairs climbed from 6,000 in 1985 to over 10,000 in 2004.

The demand for air travel in China and India has been extraordinary. Just 25 years ago, there was a total of 180 air routes in China (18 international) and only 3.43 million passengers took to the vacant skies each year. Incredibly in 2004, China overtook Japan as the largest air travel market in Asia and is second only to the US in terms of total scheduled departing seats. China is now experiencing growth rates of 15% a year. In 2004, passenger traffic soared 38% to 120 million passengers and cargo/mail carriage climbed 24.5% to 2.7 million tons."

World Airline Traffic Table 20.1
1950 – 2005

	Passengers	Increase	(Millions)	Increase	(millions)	Increase	Comment
1950	31,000,000		28,000		730		
1955	68,000,000	119.35%	61,000	117.86%	1,240	69.88%	*1
1960	106,000,000	55.88%	109,000	78.69%	2,040	64.52%	
1965	177,000,000	66.98%	198,000	81.65%	4,800	135.29%	*2
1970	382,954,000	116.36%	460,481	132.57%	12,018	150.38%	*3
1975	534,024,000	39.45%	697,285	51.43%	19,371	61.18%	*4
1980	748,288,000	40.12%	1,089,128	56.20%	29,376	51.65%	
1985	899,218,000	20.17%	1,367,347	25.55%	39,837	35.61%	*5
1990	1,165,156,000	29.57%	1,894,245	38.53%	58,796	47.59%	
2000	1,656,283,000	27.05%	3,017,354	34.21%	118,080	42.04%	
2005	2,022,000,000	22.08%	3,719,700	23.28%	142,579	20.75%	*7

*1 Economy Class introduced to the North Atlantic in 1952 and spread globally
*2 Jet aircraft start to have an impact
*3 Widespread use of jets
*4 Recession plus oil crisis
*5 Recession
*6 Iraq War
*7 911. Iraq War and SARS
RPK** Revenue Passenger Kilometers (Numbers of passengers x distance traveled)
FTK** Freight Ton Kilometers (Tons of cargo X distance carried)

To handle expected growth McDonnell Douglas tried to launch a super jumbo-the MD-12 - in the early 1990s with a group of Asian aerospace companies but market response was lukewarm.

THE ROLE OF HUBS

Over the past ten years, one of the major differences between Boeing and Airbus has been the way they have viewed the role of hubs in the future growth of air travel.

Broadly, Airbus has held the view that hubs would play a far greater role with passengers being flown on its 555-seat A380, while Boeing has maintained that passengers would rather go non-stop on mid-size jets like its 250 to 290 seat 787 and its 300 to 365 seat 777.

Naturally, as the market grows in size, more cities will be connected either directly or through major hubs. While passengers may wish to go non-stop, sometimes it is more practical and economical to route passengers through hubs.

Hubs have evolved for a number of important reasons:

- They act as a consolidation point for passengers from smaller cities so a wider range of city-pair services can be offered to passengers.
- Historically, aircraft did not have the range to allow non-stops between many destinations. Up until 1995 the only aircraft that could fly from Singapore to London with a full payload was the 747-400.
- Airlines used hubs – usually their home base – to control traffic, particularly in the US.
- Use of larger aircraft through hubs can reduce fares, depending on landing changes, taxes and congestion.
- Both domestic and international markets were regulated preventing airlines from starting new services.

One of the major problems for hubs is that as they get bigger, the connections can become more difficult with delays more likely as passengers become lost and miss connections. Also, the often touted cost benefits of using larger aircraft through hubs reduces as the significant landing costs associated with major hubs are undercut by regional airports. In March 2004, landing a 747-400 at Japan's Narita Airport cost $8,952 compared to just $100 for landing a 737-800 at Perth, Western Australia.

According to highly respected industry author Professor Nawal Taneja, in his latest book (*FASTEN YOUR SEAT BELT: The Customer is Flying the Plane*), major US airlines are starting to abandon their traditional hub models to win traffic back. Taneja says some airlines are closing hubs and redesigning others to smooth out peaks and troughs in schedules.

Increased performance of the Boeing 737 series has been instrumental in introducing non-stops in the US and Europe and more recently in Asia.
Misha Popov

THE POWER OF NON-STOPS

"We focus on non-stops. It's what the customer wants and it's a lot cheaper to fly non-stop." – Gary Kelly, CEO Southwest Airlines

There can no more emphatic endorsement of the non-stop argument than that of Kelly's when he was interviewed by trade journal *Air Transport World's* (*ATW*) Editor in Chief Perry Flint in April 2005.

Texas-based Southwest Airlines created the blueprint for the Low Cost Carrier (LCC) revolution that has transformed air travel over the past 20 years. In April 2005, the airline had 417 Boeing 737s operating 2,900 flights a day to 59 destinations in 31 states and has been profitable in 31 of the last 32 years. The success of Southwest Airlines and other LCCs, such as JetBlue, has seen the LCC share of the US domestic market climb from 7% in 1990 to 25% in 2004, according to the European Low Fare Airline Association (ELFAA) report Benefits of Low Fare Airlines 2004.

JetBlue, based in New York, flies non-stop to many secondary airports. An excellent example is the Los Angeles area where the airline operates a host of daily non-stops to secondary airports like Long Beach, Ontario and Burbank rather than serving Los Angeles International Airport (LAX). The airline has virtually no hub operation. Reinforcing the popularity of non-stops on JetBlue's website, in May 2005, the airline touted five new non-stop services to start during that month.

In Europe, the message is the same with LCC giant Ryanair and its fleet of Boeing 737s. Founder and CEO Michael O'Leary told *ATW* in April 2004 that Ryanair was "strictly a point-to-point operator and shivered at the thought of a hub operation." The airline has 12 bases and the key to the airline's runaway success is its strategy of flying to uncongested low cost airports. According to an ELFAA report in 2004, "the low cost model is dependent on direct point-to-point flights, no transfers and short haul flights." The report cites the European Commission, which reported in 2004 that there had been a 30% increase in new routes in Europe since 1993.

ETOPS AND LIBERALIZATION OPEN UP AIR ROUTES

Since the medium-range twin-engine 250-seat Boeing 767 pioneered Extended Twin Operations (ETOPS) across the Atlantic in February 1985, there has been a proliferation of non-stop services in that market and around the globe. ETOPS, which was covered in Chapter 5, has transformed the dynamics of the airline industry as has market liberalization as more and more countries opt for "open skies" agreements.

Not so long ago, Tokyo's Narita was dubbed "747 city" but when ETOPS requirements were eased to 207-minutes in the late 1990s, the airport became home to smaller twin engine types such as the 777-200ER, A330s and 777-300ERs.

According to Official Airline Guide (OAG) figures in August 2004, the 747's share of departures at Narita had plummeted from nearly 80% to just 40% in seven years. This drop was accelerated by the opening of an additional runway freeing up landing and departure slot constraints.

In Hong Kong, the story is similar. The 747s had a 45% share when Hong Kong's airport was based at Kai Tak with its single runway and its spectacular but difficult approach. Since the opening of Hong Kong International Airport with its parallel runways, the 747's dominance has waned and now accounts for only 18% of departures.

OAG figures taken in the month of August 2005, excluding regional jets and turbo-props, are emphatic on the development of the market over the past 20 years.

They show:

- Non-stops routes have jumped from 5,700 to over 10,400.
- Frequency of service has almost tripled.
- Average size of aircraft has declined slightly.

According to an InterVISTAS-ga2 report The Economic Impact of Air Service Liberalization, "liberalizing only 320 bilateral agreements of the existing thousands would create 24.1 million full-time jobs and generate an additional $490 billion in Gross Domestic Product."

This study found:

- Traffic growth subsequent to liberalization of air services agreements between countries typically averaged between 12% and 35% -- significantly greater than during the years preceding liberalization. In a number of situations, growth was at rates exceeding 50%, and in some cases, reached almost 100% of the pre-liberalization rates.
- A simulation of the likely results of liberalizing 320 country pair markets that are not today in an "open skies" (deregulated) mode indicate traffic growth, on average, of almost 63%. This is substantially higher than typical world traffic growth of around 6 to 8%.
- The creation of the Single European Aviation Market in 1993 led to an average annual growth rate in traffic between 1995 and 2004 that was almost double the rate of growth in the years 1990 to 1994. This produced about 1.4 million new jobs.
- An examination of 190 countries and 2,000 bilateral air service agreements reveal that there are still a number of countries that place a priority on protecting their flag carrier(s), rather than enhancing the overall welfare of the broader public interest.

Four A380s in formation. *Airbus*

GIANTS FOR SLOT LIMITED AIRPORTS

At holiday peak times, most airlines would just love a 2,000-passenger aircraft to move all the travelers who want to get home for the holiday season. Problem is that such an aircraft would be useless for the rest of the year.

But for airlines that are restricted in the number of slots they can obtain into airports such as Heathrow and LAX, an aircraft larger than the 747-400 makes sense. Airlines such as Emirates, Qantas and Singapore Airlines are all "slot limited" into Heathrow, an airport critical to their marketing strategies, and all were among the first to order the 555-seat Airbus A380.

Boeing has responded to the requirement for larger aircraft by launching a new version of its 747, dubbed the 747-8, which will seat up to 479 passengers. The aircraft combines a lengthened fuselage with a radical upgrade of the wing and uses 787 engines. It is covered in more detail on page 228.

Boeing does not believe that the market for the A380 will be as great as Airbus' prediction of 1,262 up to the year 2023. Boeing's view is supported by one of the world's most influential aircraft lessors International Lease Finance Corporation's (ILFC) founder and CEO Steven Udvar-Hazy, who told *The Seattle Times* at the International Society of Transport Aircraft Trading (ISTAT) conference on 29 March 2006 that Airbus would only sell "at best 300 to 400" A380s.

Boeing believes that the market size for aircraft of the 747 size and above is 790 units through 2023. The US manufacturer suggests that 400 of these units will be 400 to 500 seat aircraft (where its 747-8 is targeted) and the balance will have more than 500 seats.

Airbus argues however, that Asia/Pacific population concentrations mean that long-haul flights to the region will not become fragmented like trans-Atlantic operations, and therefore, airlines will need greater numbers of bigger aircraft.

And while it is true that there are a number of heavy population concentrations, such as Singapore and Hong Kong, it is also true, argues Boeing, that there are many mature population centers in North America and Europe that want non-stop flights to Asia. So the market will fragment at one end only.

Balancing that, Boeing points out that, unlike many US and European airports, the major hubs in Asia have extremely modern airports with plenty of runway capacity to handle more flights – and they have aggressive governments that want to attract more services.

An excellent example of this argument is Japan, where in the past Japan Airlines (JAL) and All Nippon Airways (ANA) have used all economy 747s with up to 563 seats for domestic flights. However, with the advent of the more economical 777-300 with 490 seats in a similar configuration and an additional runway at Narita (and a fourth at Haneda), both airlines are phasing out some of their domestic 747s. Japan Airlines says that "with a 40% increase in takeoff and landing slots at Haneda (from 2008), JAL will improve customer convenience, enhance operational efficiency through a better match of supply to demand and expand its network with the 777 and the 787."

According to the world's largest airfreight company FedEx, aircraft carry around 2% of international trade by volume, but around 40% by value.

And that 40% is made up of some extraordinary cargo. Live lobsters from Perth, Western Australia to Tokyo and Seattle, flowers from Hawaii to Paris, strawberries from Sydney to New York, but that depends on the Californian crop – and the list goes on. Computers, bees, racing cars, newspapers, magazines and chilled tuna are all carried by air and usually overnight while we sleep.

Boeing and Airbus suggest that world air cargo growth is expected to expand at an average annual rate of 6.1% over the next twenty years with markets linked to Asia leading the way.

According to Boeing's VP of Marketing and Business Strategy Nicole Piasecki, Boeing expects that by 2025, "766 new production freighters will be required valued at $170 billion." In fact, Boeing forecasts that the number of cargo aircraft in service worldwide will grow to 3,563 by 2025 from 1,789 in 2005.

Just part of FedEx's major cargo hub at Nashville, USA. *FedEx*

And while the number of airplanes will increase, so will the size. Wide-body freighters – such as MD-11Fs, 747s and A300s – are currently 44% of the fleet and the percentage is expected to grow to 60% of the fleet.

In the 20 months from January 2005 to September 2006, Boeing for instance had sold 89 747-400Fs and 747-8Fs. And many of those 747 freighters were going to China and Asia where growth rates were expected to be higher than in the past at 10.8% and 8.6% a year respectively.

But as well as new airplanes, many will be modified passenger planes. In fact, Boeing predicted that three-quarters of freighter fleet additions during the next 20 years will come from modified passenger and combi-airplanes, satisfying both market growth and replacement needs.

One such aircraft was the Boeing 747-400BCF (Boeing Converted Freighter) which was launched in 2004 by Cathay Pacific Airways. The first aircraft flew in early October 2005 and was redelivered to Cathay Pacific on December 19, 2006.

Cargolux Airlines was a launch customer for the Boeing 747-8F in November 2005. *Boeing*

To meet the demand for new freighters, Boeing launched the 777-200F and 747-8 in 2005. Based on the record-breaking 777-200LR, the 777-200F will carry 229,000lbs (103.9 metric tons) a distance of 4,895nm (9,065km) at maximum takeoff weight of 766,000lbs (347,450kg), making it the world's longest-range freighter.

The 747-8F is 18.3ft (5.6m) longer than the standard 747 and has a structural payload capability of 154 tons (140 metric tons) with a range of 4,475nm (8,275km). Powered by 787 GEnx engines, the 747-8 will have nearly equivalent trip costs and 14% lower ton-mile costs than the 747-400, plus 16% more revenue cargo volume than its predecessor.

Boeing has radically modified three Boeing 747s to transport fuselage sections of its new Boeing 787 Dreamliner. Here the first sections from Japan arrive at Charleston, South Carolina for final integration. *Boeing*

Spectacular growth is expected to make the skies even more crowded. Here a Northwest 747-400 turns above a Singapore Airlines 777.
Captain Kevin Tate

BOOMING MARKET

Both Airbus and Boeing have agreed that the growth in air travel between 2006 and 2025 will be extraordinary. Whether it is to visit friends and relatives, enjoy a holiday or for business in another city, air travel makes it possible. Air travel is booming as airfares decline in comparison to average weekly earnings.

In Boeing's Current Market Outlook 2006 forecast, it said that the replacement and growth markets will require a total of 27,210 new airplanes over 100 seats in size, worth $2.6 trillion over the next 20 years.

The numbers are impressive. At the end of 2005, there were 17,300 commercial jet aircraft in service and Boeing forecast that over the next twenty years 9,580 would be required to replace older airplanes while 17,630 would be needed to handle market growth. In addition, 2,220 of the existing fleet would be converted to freighters giving a world fleet total of 35,970 airplanes.

Fueling that demand for aircraft are passenger and cargo growth rates that are almost identical to what we have seen in the 20 years prior to 2005. World Gross Domestic Product has increased 2.9% a year from 1985 to 2005 and Boeing forecasts a 3.1% growth in the next twenty years driven a little faster by China's and India's growth.

Air travel and air cargo growth is always higher than GDP and

An-225 is the world's largest freighter.

Boeing suggests a 4.9% growth in passengers per year and a 6.1% growth in cargo. Airbus's numbers are almost identical. However as discussed on page 226, Airbus forecast in 2004 that to handle the growth, passenger aircraft will get larger rather than smaller and more numerous.

Airbus said that "to accommodate this three-fold growth in passenger traffic, the number of flights offered on passenger routes and the number of passenger aircraft in service will more than double in 20 years, accompanied by the use of larger aircraft. Airbus therefore forecasts that the average number of seats per passenger aircraft will increase by 20 per cent from 181 to 215 over this period."

However, in recent times the market appeared to be moving in a direction more congruent with the Boeing view with the orders for its mid-size 230 to 290-seat 787 outstripping orders for the 555-seat A380 at a ratio of 3:1. Those 787 orders make the aircraft the fastest selling commercial aircraft in history.

And air travel fuels one of the world's fastest growing and largest industries, Tourism and Travel – a fact lost to many politicians, who consider it a "Mickey Mouse" industry.

The numbers are extraordinary. In 2006, according to the London based World Tourism and Travel Council, the world's tourism and travel industry will generate:

- $6,477.2 billion of economic activity
- 10.3% of total GDP
- 234,305,000 jobs or 8.7% of total world employment
- Capital investment of $1,010.7 billion or 9.3% of total investment

EasyJet Boeing 737-700 over Germany taken from an Emirates Boeing 777-200.

YEAR	Yield (in current cents)			U.S. CPI (1982-84 =100)	System Yield (in 1978 cents)
	Domestic	Internat'l	System		
1938	5.18	8.34	5.50	14.40	25.41
1939	5.10	8.57	5.43	13.90	25.48
1940	5.07	8.83	5.39	14.00	25.11
1941	5.04	8.61	5.42	14.70	24.02
1942	5.34	8.86	5.85	16.30	23.39
1943	5.35	7.94	5.69	17.30	21.43
1944	5.34	7.83	5.65	17.60	20.95
1945	4.95	8.68	5.39	18.00	19.52
1946	4.63	8.31	5.21	19.50	17.41
1947	5.05	7.77	5.67	22.30	16.58
1948	5.76	8.01	6.30	24.10	17.05
1949	5.78	7.72	6.23	23.80	17.07
1950	5.56	7.28	5.94	24.10	16.06
1951	5.61	7.1	5.91	26.00	14.81
1952	5.57	7.01	5.85	26.50	14.39
1953	5.46	6.84	5.72	26.70	13.96
1954	5.41	6.76	5.66	26.90	13.72
1955	5.36	6.66	5.60	26.80	13.63
1956	5.33	6.68	5.58	27.20	13.38
1957	5.31	6.55	5.54	28.10	12.86
1958	5.64	6.46	5.80	28.90	13.09
1959	5.88	6.29	5.96	29.10	13.35
1960	6.09	6.35	6.14	29.60	13.53
1961	6.28	6.08	6.24	29.90	13.60
1962	6.45	5.87	6.31	30.20	13.63
1963	6.17	5.82	6.09	30.60	12.98
1964	6.12	5.45	5.95	31.00	12.52
1965	6.06	5.29	5.87	31.50	12.14
1966	5.83	5.16	5.67	32.40	11.41
1967	5.64	5.01	5.49	33.40	10.73
1968	5.61	4.95	5.46	34.80	10.23
1969	5.79	5.18	5.68	36.70	10.09
1970	6.00	5.01	5.79	38.80	9.73
1971	6.33	5.08	6.06	40.50	9.76
1972	6.40	4.98	6.08	41.80	9.49
1973	6.63	5.32	6.34	44.40	9.32
1974	7.52	6.39	7.29	49.30	9.64
1975	7.69	7.17	7.59	53.80	9.20
1976	8.16	7.15	7.97	56.90	9.13
1977	8.61	7.61	8.42	60.60	9.06
1978	8.49	7.49	8.29	65.20	8.29
1979	8.96	7.66	8.70	72.60	7.81
1980	11.49	8.79	10.99	82.40	8.70
1981	12.74	9.47	12.34	90.90	8.85
1982	12.02	9.57	11.77	96.50	7.95
1983	12.05	9.76	11.62	99.60	7.61
1984	12.80	9.38	12.11	103.90	7.60
1985	12.21	9.27	11.66	107.60	7.07
1986	11.08	9.63	10.93	109.60	6.50
1987	11.45	9.74	11.11	113.60	6.38
1988	12.31	10.40	11.88	118.30	6.55
1989	13.08	10.36	12.43	124.00	6.54
1990	13.43	10.83	12.76	130.70	6.37
1991	13.24	11.32	12.74	136.20	6.10
1992	12.85	11.56	12.51	140.30	5.81
1993	13.74	11.28	13.13	144.50	5.89
1994	13.12	11.18	12.65	148.20	5.54
1995	13.52	11.13	12.92	152.40	5.51
1996	13.76	10.92	13.05	156.90	5.41
1997	13.97	10.96	13.18	160.50	5.35
1998	14.08	10.38	13.11	163.00	5.24
1999	13.96	10.06	12.94	166.60	5.06
2000	14.57	10.59	13.51	172.20	5.12
2001	13.41	9.65	12.42	177.10	4.57
2002	11.98	9.95	11.46	179.90	4.15
2003	12.29	10.14	11.78	184.00	4.17
2004	12.03	10.60	11.67	188.90	4.03
2005	12.28	11.16	12.00	195.30	4.00

Source: Air Transport Association

US Annual Passenger Prices (Yield)

This table gives the fares paid by passengers in cents per mile in both then year dollars and expressed in 1978 cents. For further comparison the US CPI is added with an index in 1982-84 of 100).

WORLD SCHEDULED AIRLINES: SYSTEM SCHEDULED TRAFFIC AND OPERATIONS, 1939-2005
(prior to 1970, results exclude CIS/USSR; for all years, U.S. data reflects majors and nationals only)

	AIRCRAFT VOLUMES			PASSENGER VOLUMES				Freight	
YEAR	Kms (mils)	Departures (thous)	Hours (thous)	Psrgrs (thous)	RPKs (mils)	ASKs (mils)	PLF (%)	Tons (000)	RTKs (millions)
1939	295	N/A	N/A	N/A	2,030	N/A	N/A	N/A	N/A
1940	300	N/A	N/A	N/A	2,530	N/A	N/A	N/A	N/A
1941	340	N/A	N/A	N/A	3,280	N/A	N/A	N/A	N/A
1942	320	N/A	N/A	N/A	3,580	N/A	N/A	N/A	N/A
1943	325	N/A	N/A	N/A	4,265	N/A	N/A	N/A	N/A
1944	415	N/A	N/A	N/A	5,490	N/A	N/A	N/A	N/A
1945	600	N/A	2,500	9,000	8,000	N/A	N/A	N/A	240
1946	940	N/A	3,800	18,000	16,000	N/A	N/A	N/A	220
1947	1,140	N/A	4,200	21,000	19,000	N/A	N/A	N/A	400
1948	1,270	N/A	4,600	24,000	21,000	N/A	N/A	N/A	590
1949	1,350	N/A	4,800	27,000	24,000	N/A	N/A	N/A	760
1950	1,440	N/A	5,000	31,000	28,000	46,000	60.9	N/A	930
1951	1,620	N/A	5,700	42,000	35,000	55,000	63.6	N/A	1,100
1952	1,780	N/A	6,100	46,000	40,000	64,000	62.5	N/A	1,200
1953	1,950	N/A	6,500	53,000	47,000	75,000	62.7	N/A	1,270
1954	2,060	N/A	6,700	59,000	52,000	86,000	60.5	N/A	1,370
1955	2,290	N/A	7,300	68,000	61,000	99,000	61.6	N/A	1,610
1956	2,540	N/A	8,000	77,000	71,000	114,000	62.3	N/A	1,800
1957	2,840	N/A	8,700	86,000	82,000	133,000	61.7	N/A	1,960
1958	2,930	N/A	8,800	88,000	85,000	146,000	58.2	N/A	2,040
1959	3,090	N/A	9,000	98,000	98,000	162,000	60.5	N/A	2,350
1960	3,110	6,600	8,600	106,000	109,000	184,000	59.2	N/A	2,650
1961	3,120	6,600	8,000	111,000	117,000	212,000	55.2	N/A	3,080
1962	3,240	6,600	7,800	121,000	130,000	243,000	53.5	N/A	3,580
1963	3,430	6,700	7,900	135,000	147,000	274,000	53.6	N/A	3,970
1964	3,700	7,100	8,200	155,000	171,000	306,000	55.9	N/A	4,670
1965	4,100	7,500	8,700	177,000	198,000	354,000	55.9	N/A	5,900
1966	4,480	7,800	9,300	200,000	229,000	397,000	57.7	N/A	7,230
1967	5,280	8,600	10,200	233,000	273,000	479,000	57.0	N/A	8,420
1968	5,993	9,074	10,992	260,277	309,422	578,885	53.5	N/A	10,503
1969	6,704	9,473	11,761	293,230	350,899	674,973	52.0	3,970	12,286
1970	7,004	9,486	12,036	382,954	460,481	839,930	54.8	6,083	15,087
1971	7,054	9,546	12,016	410,919	494,137	914,070	54.1	6,673	16,125
1972	7,209	9,646	12,192	450,136	560,078	980,766	57.1	7,284	17,801
1973	7,520	9,889	12,716	488,514	618,184	1,072,599	57.6	8,223	20,408
1974	7,375	9,613	12,417	514,496	656,426	1,107,532	59.3	8,651	21,900
1975	7,516	9,683	12,580	534,024	697,285	1,178,880	59.1	8,680	22,270
1976	7,840	9,986	13,028	576,421	763,762	1,269,815	60.1	9,336	24,571
1977	8,089	10,118	13,345	610,322	818,300	1,346,260	60.8	10,046	26,805
1978	8,498	10,379	13,985	678,645	936,352	1,451,184	64.5	10,625	29,205
1979	9,147	10,666	14,887	754,117	1,060,236	1,607,188	66.0	10,996	31,436
1980	9,350	10,691	15,112	748,288	1,089,128	1,723,903	63.2	11,089	33,057
1981	9,113	10,250	14,749	752,271	1,119,066	1,756,539	63.7	10,915	34,675
1982	9,140	10,379	14,796	765,806	1,142,193	1,794,646	63.6	11,568	35,413
1983	9,395	10,820	15,360	797,810	1,189,767	1,852,088	64.2	12,257	39,112
1984	10,102	11,453	16,492	847,931	1,278,176	1,972,296	64.8	13,416	43,978
1985	10,958	11,952	17,223	899,218	1,367,347	2,081,018	65.7	13,724	44,236
1986	11,491	12,698	18,646	960,012	1,452,055	2,234,730	65.0	14,687	47,735
1987	12,266	13,306	19,902	1,027,856	1,589,467	2,367,532	67.1	16,058	53,021
1988	13,017	13,942	21,269	1,082,474	1,705,432	2,524,094	67.6	17,215	58,099
1989	13,493	-13,945	22,815	1,109,478	1,773,703	2,608,046	68.0	18,088	62,202
1990	14,371	14,661	23,389	1,165,156	1,894,245	2,800,844	67.6	18,423	64,121
1991	14,262	14,269	23,316	1,135,185	1,845,418	2,779,494	66.4	17,465	63,630
1992	15,690	14,819	24,851	1,145,553	1,928,922	2,930,185	65.8	17,647	67,761
1993	17,118	15,777	27,322	1,142,382	1,949,421	3,013,411	64.7	18,053	73,671
1994	18,249	17,038	29,161	1,233,341	2,099,936	3,169,342	66.3	20,522	82,626
1995	19,470	17,816	30,868	1,303,645	2,248,215	3,358,601	66.9	22,189	88,765
1996	20,601	18,758	32,836	1,391,085	2,431,695	3,563,774	68.2	23,234	94,996
1997	21,630	19,320	34,350	1,456,690	2,573,010	3,727,900	69.0	26,360	108,870
1998	22,438	19,686	35,559	1,471,470	2,628,116	3,837,725	68.5	26,496	107,575
1999	23,742	20,739	37,683	1,562,324	2,797,803	4,050,783	69.1	28,103	114,373
2000	25,174	21,325	39,763	1,656,283	3,017,354	4,258,619	70.9	30,370	124,130
2001	25,612	21,134	40,777	1,640,277	2,949,553	4,271,858	69.0	28,829	116,103
2002	25,418	21,705	40,099	1,638,567	2,964,531	4,167,106	71.1	31,420	124,412
2003	26,264	22,093	41,196	1,691,233	3,019,104	4,227,863	71.4	33,522	130,291
2004	29,163	23,754	45,441	1,888,403	3,445,324	4,704,729	73.2	36,661	143,612
2005	30,845	24,904	48,019	2,022,259	3,719,703	4,973,159	74.8	37,660	147,240

ANNUAL TRAFFIC AND CAPACITY
U.S. Scheduled Airlines — Scheduled Services

YEAR	Revenue Aircraft Departures (000)	Revenue Passengers Enplaned (000)			Revenue Passenger Miles (millions)			Available Seat Miles (millions)			Passenger Load Factor (percent)		
	System	Domestic	Internat'l	System	Domestic	Internat'l	System	Domestic	Internat'l	System	Domestic	Internat'l	System
1939	N/A	1,735	129	1,864	683	72	755	1,215	134	1,350	56.2	53.5	55.9
1940	N/A	2,803	163	2,966	1,052	100	1,152	1,817	175	1,993	57.9	56.9	57.8
1941	N/A	3,849	229	4,078	1,385	163	1,548	2,342	249	2,591	59.1	65.4	59.7
1942	N/A	3,129	277	3,406	1,418	237	1,655	1,963	314	2,277	72.2	75.7	72.7
1943	N/A	3,012	303	3,315	1,632	246	1,878	1,855	310	2,164	88.0	79.4	86.8
1944	N/A	4,027	361	4,388	2,177	312	2,489	2,435	393	2,828	89.4	79.3	88.0
1945	N/A	6,541	511	7,052	3,360	450	3,810	3,811	588	4,399	88.2	76.6	86.6
1946	N/A	12,164	1,091	13,255	5,945	1,104	7,049	7,550	1,561	9,110	78.7	70.7	77.4
1947	N/A	12,822	1,428	14,250	6,105	1,814	7,920	9,364	2,934	12,298	65.2	61.8	64.4
1948	2,092	13,094	1,447	14,541	5,997	1,894	7,890	10,417	3,303	13,720	57.6	57.3	57.5
1949	2,262	15,121	1,599	16,720	6,768	2,060	8,827	11,712	3,639	15,351	57.8	56.6	57.5
1950	2,457	17,468	1,752	19,220	8,029	2,214	10,243	13,125	3,717	16,842	61.2	59.6	60.8
1951	2,596	22,711	2,140	24,851	10,590	2,614	13,204	15,615	4,369	19,984	67.8	59.8	66.1
1952	2,737	25,176	2,391	27,567	12,559	3,065	15,624	19,170	4,955	24,125	65.5	61.9	64.8
1953	2,960	28,901	2,745	31,646	14,794	3,451	18,245	23,337	5,624	28,962	63.4	61.4	63.0
1954	3,002	32,529	2,919	35,448	16,802	3,810	20,613	26,922	6,455	33,377	62.4	59.0	61.8
1955	3,281	38,221	3,488	41,709	19,852	4,499	24,351	31,371	7,203	38,574	63.3	62.5	63.1
1956	3,503	41,937	4,068	46,005	22,399	5,226	27,625	35,366	8,308	43,674	63.3	62.9	63.3
1957	3,771	45,162	4,304	49,466	25,379	5,882	31,261	41,746	9,313	51,059	60.8	63.2	61.2
1958	3,629	48,298	4,772	53,070	25,375	6,124	31,499	42,724	10,392	53,115	59.4	58.9	59.3
1959	3,910	54,958	5,338	60,296	29,308	7,064	36,372	48,405	10,842	59,247	60.5	65.2	61.4
1960	3,853	56,351	5,906	62,257	30,557	8,306	38,863	52,220	13,347	65,567	58.5	62.2	59.3
1961	3,750	56,900	6,112	63,012	31,062	8,769	39,831	56,087	15,770	71,857	55.4	55.6	55.4
1962	3,660	60,739	7,079	67,818	33,623	10,138	43,760	63,888	18,724	82,612	52.6	54.1	53.0
1963	3,788	69,366	8,037	77,403	38,457	11,905	50,362	72,255	22,590	94,845	53.2	52.7	53.1
1964	3,955	79,139	9,381	88,520	44,141	14,352	58,494	80,524	25,791	106,316	54.8	55.6	55.0
1965	4,198	92,073	10,847	102,920	51,887	16,789	68,676	94,787	29,533	124,320	54.7	56.8	55.2
1966	4,374	105,789	12,272	118,061	60,591	19,298	79,889	104,669	33,176	137,844	57.9	59.2	58.0
1967	4,946	128,479	14,020	142,499	75,487	23,259	98,747	133,700	41,119	174,819	56.5	56.6	56.5
1968	5,348	145,774	16,407	162,181	87,508	26,451	113,958	166,871	49,575	216,446	52.4	53.4	52.6
1969	5,378	158,405	13,493	171,898	102,717	22,703	125,420	206,434	44,412	250,846	49.8	51.1	50.0
1970	5,120	153,662	16,260	169,922	104,147	27,563	131,710	213,160	51,960	265,120	48.9	53.0	49.7
1971	4,999	156,195	17,474	173,669	106,438	29,219	135,658	221,503	58,320	279,823	48.1	50.1	48.5
1972	5,046	172,452	18,897	191,349	118,138	34,268	152,406	226,614	60,797	287,411	52.1	56.4	53.0
1973	5,135	183,272	18,936	202,208	126,317	35,640	161,957	244,699	65,898	310,597	51.6	54.1	52.1
1974	4,726	189,733	17,725	207,458	129,732	33,186	162,919	233,880	63,126	297,006	55.5	52.6	54.9
1975	4,705	188,746	16,316	205,062	131,728	31,082	162,810	241,282	61,724	303,006	54.6	50.4	53.7
1976	4,833	206,279	17,039	223,318	145,271	33,717	178,988	261,248	61,574	322,822	55.6	54.8	55.4
1977	4,937	222,283	18,043	240,326	156,609	36,610	193,219	280,619	64,947	345,566	55.8	56.4	55.9
1978	5,016	253,957	20,759	274,716	182,669	44,112	226,781	299,542	69,209	368,751	61.0	63.7	61.5
1979	5,400	292,700	24,163	316,863	208,891	53,132	262,023	332,796	83,330	416,126	62.8	63.8	63.0
1980	5,353	272,829	24,074	296,903	200,829	54,363	255,192	346,028	86,507	432,535	58.0	62.8	59.0
1981	5,212	265,304	20,672	285,976	198,715	50,173	248,888	346,172	78,725	424,897	57.4	63.7	58.6
1982	4,964	274,342	19,760	294,102	210,149	49,495	259,644	359,528	80,591	440,119	58.5	61.4	59.0
1983	5,034	296,721	21,917	318,638	226,909	54,920	281,829	379,150	85,388	464,538	59.8	64.3	60.7
1984	5,448	321,047	23,636	344,693	243,692	61,424	305,116	422,507	92,817	515,323	57.7	66.2	59.2
1985	5,835	357,109	24,913	382,022	270,584	65,819	336,403	445,826	101,963	547,788	60.7	64.6	61.4
1986	6,427	393,864	25,082	418,946	302,090	64,456	366,546	497,991	109,445	607,436	60.7	58.9	60.3
1987	6,581	416,831	30,847	447,678	324,637	79,834	404,471	526,958	121,763	648,721	61.6	65.6	62.3
1988	6,700	419,210	35,404	454,614	329,309	93,992	423,302	536,663	140,140	676,802	61.4	67.1	62.5
1989	6,622	416,331	37,361	453,692	329,975	102,739	432,714	530,079	154,297	684,376	62.3	66.6	63.2
1990	6,924	423,565	41,995	465,560	340,231	117,695	457,926	563,065	170,310	733,375	60.4	69.1	62.4
1991	6,783	412,360	39,941	452,301	332,556	115,389	447,955	543,638	171,561	715,199	61.2	67.3	62.6
1992	7,051	431,693	43,415	475,108	347,931	130,622	478,554	557,989	194,784	752,722	62.4	67.1	63.6
1993	7,245	443,172	45,348	488,520	354,177	135,508	489,684	571,489	200,151	771,641	62.0	67.7	63.5
1994	7,531	481,755	47,093	528,848	378,990	140,391	519,382	585,438	198,893	784,331	64.7	70.6	66.2
1995	8,062	499,000	48,773	547,773	394,708	145,948	540,656	603,917	203,160	807,078	65.4	71.8	67.0
1996	8,230	530,708	50,526	581,234	425,596	153,067	578,663	626,389	208,682	835,071	67.9	73.3	69.3
1997	8,127	542,001	52,724	594,725	442,640	160,779	603,419	640,319	216,913	857,232	69.1	74.1	70.4
1998	8,292	559,653	53,232	612,885	454,430	163,656	618,087	649,362	224,728	874,089	70.0	72.8	70.7
1999	8,627	582,880	53,079	635,959	480,134	171,913	652,047	687,502	230,917	918,419	69.8	74.4	71.0
2000	9,035	610,600	55,550	666,150	508,403	184,354	692,757	714,454	242,496	956,950	71.2	76.0	72.4
2001	8,788	570,126	52,003	622,129	480,348	171,352	651,700	695,200	235,311	930,511	69.1	72.8	70.0
2002	9,187	560,107	52,769	612,877	476,044	165,098	641,102	676,949	215,606	892,554	70.3	76.6	71.8
2003	10,839	592,412	53,863	646,276	500,271	156,638	656,909	689,069	204,755	893,824	72.6	76.5	73.5
2004	11,401	640,698	62,222	702,921	551,937	181,743	733,680	741,677	229,788	971,466	74.4	79.1	75.5
2005	11,562	670,418	68,210	738,629	579,690	199,324	779,014	752,482	250,854	1,003,336	77	79.5	77.6
2006	11,254	671,606	72,847	744,454	585,378	212,030	797,408	740,924	265,437	1,006,361	79	79.9	79.2

THE WORLD'S TOP 25 AIRLINES in OPERATING REVENUE

2000			2001			2002			2003			2004			2005		
Rank	Airline	Op. Revenue (000)	Rank	Airline	Op. Revenue (000)	Rank	Airline	Op. Revenue (000)	Rank	Airline	Op. Revenue (000)	Rank	Airline	Op. Revenue (000)	Rank	Airline	Op. Revenue (000)
1	American	19,703,000	1	American	18,963,000	1	Lufthansa Group	18,057,000	1	Lufthansa Group	21,480,000	1	Lufthansa Group	25,655,000	1	Air France-KLM	25,901,248
2	UAL Corp	19,352,000	2	UAL Group	16,138,000	2	AMR Corp	17,299,000	2	AMR Corp	17,440,000	2	Air France-KLM	24,641,145	2	Lufthansa Group	21,397,090
3	Delta	16,741,000	3	FedEx	15,166,933	3	JAL System	17,105,371	3	JAL System	16,945,105	3	JAL System	19,794,387	3	AMR Group	20,712,000
4	FedEx	15,596,508	4	Lufthansa Group	14,687,000	4	FedEx	15,941,157	4	FedEx	16,806,533	4	FedEx	18,675,535	4	FedEx	20,712,000
5	JAL Group	13,487,000	5	Delta	13,879,000	5	UAL Corp	14,286,000	5	Air France Group	15,491,570	5	AMR Group	18,645,000	5	JAL Group	18,713,200
6	Lufthansa Group	13,356,240	6	JAL Group	12,373,846	6	Air France Group	13,702,000	6	Delta	14,203,030	6	United	16,391,000	6	United	17,379,000
7	British Airways	13,230,428	7	British Airways	12,176,000	7	Delta	13,305,000	7	United	13,724,000	7	Delta	15,002,000	7	Delta	16,191,000
8	Northwest	11,415,000	8	Air France Group	11,024,640	8	British Airways	12,116,500	8	British Airways	13,445,460	8	British Airways	14,680,627	8	British Airways	14,814,227
9	Air France Group	10,790,000	9	Northwest	9,905,000	9	All Nippon	10,147,000	9	Northwest	9,510,000	9	All Nippon	11,752,727	9	Northwest	12,286,000
10	All Nippon Group	10,129,000	10	All Nippon	9,265,000	10	Northwest	9,489,000	10	Continental	8,870,000	10	NorthWest	11,279,000	10	All Nippon	11,643,010
11	Swissair Group	10,076,424	11	Continental	8,969,000	11	Continental	8,402,000	11	Qantas	8,525,260	11	Continental	9,744,000	11	Continental	11,208,000
12	Continental	9,899,000	12	US Airways	8,253,356	12	SAS Group	7,429,600	12	All Nippon	8,518,230	12	SAS Group	8,830,276	12	Qantas	9,638,386
13	US Airways	9,248,000	13	Air Canada	6,040,514	13	KLM	7,004,000	13	SAS Group	7,978,000	13	Qantas	7,837,459	13	Air Canada	8,434,042
14	Air Canada	6,197,795	14	KLM	5,744,000	14	US Airways	6,977,000	14	KLM	7,379,000	14	Air Canada	7,387,000	14	SIA Group	8,236,128
15	KLM	6,115,752	15	Southwest	5,555,174	15	Air Canada	6,233,600	15	US Airways	6,846,000	15	SIA Group	7,276,093	15	SAS Group	7,776,102
16	Southwest	5,649,560	16	Singapore	5,252,977	16	SIA Group	5,930,460	16	Air Canada	6,465,786	16	US Airways	7,117,000	16	Southwest	7,584,000
17	Qantas	5,486,847	17	Qantas	5,202,000	17	Qantas	5,897,037	17	Southwest	5,936,696	17	Korean	7,031,309	17	Korean	7,486,900
18	Alitalia Group	5,146,000	18	SAS Group	4,989,000	18	Southwest	5,521,771	18	Iberia Group	5,820,354	18	Southwest	6,530,000	18	Emirates	6,597,400
19	Singapore Airlines	5,113,254	19	Alitalia Group	4,698,320	19	Korean	5,206,348	19	SIA Group	5,731,504	19	Alitalia	5,619,384	19	Cathay Pacific	6,527,000
20	SAS Group	5,054,000	20	Korean	4,284,000	20	Iberia	4,925,109	20	Alitalia	5,424,624	20	Iberia	5,549,577	20	Iberia	5,838,272
21	Cathay Pacific	4,426,194	21	Iberia	4,169,658	21	Alitalia	4,868,628	21	Korean	5,171,552	21	Emirates	5,202,100	21	Alintalia	5,681,807
22	Iberia	3,793,348	22	Cathay Pacific	3,895,808	22	Cathay Pacific	4,242,000	22	Cathay Pacific	3,792,000	22	Cathay Pacific	5,009,000	22	US Airways Group	5,077,000
23	TWA	3,584,638	23	Japan Air System	3,300,000	23	Airborne Express	3,343,736	23	Emirates	3,600,000	23	Air China	4,054,914	23	China Southern	4,746,800
24	Airborne Express	3,275,950	24	Air NZ Group	3,231,760	24	Thai	2,976,500	24	Thai	3,158,000	24	Thai	3,679,259	24	Air China	4,746,548
25	Korean Air	3,089,271	25	Airborne Express	3,211,089	25	UPS	2,851,943	25	UPS	3,046,427	25	UPS	3,349,816	25	UPS	4,105,212

THE WORLD'S TOP 25 AIRLINES in RPKs

Rank	2000 Airline	2000 No. of RPKs (000, 000)	Rank	2001 Airline	2001 No. of RPKs (000, 000)	Rank	2002 Airline	2002 No. of RPKs (000, 000)
1	United	204,235	1	United	187,666	1	American	195,812
2	American	187,60	2	American	170,883	2	United	176,152
3	Delta	173,486	3	Delta	163,663	3	Delta	152,661
4	Northwest	127,317	4	Northwest	117,658	4	Northwest	115,913
5	British Airways	118,890	5	British Airways	106,270	5	British Airways	99,710
6	Continental	103,235	6	Continental	98,374	6	Air France	96,802
7	Air France	91,801	7	Air France	94,415	7	Continental	95,510
8	Japan Airlines	88,999	8	Lufthansa	86,695	8	Lufthansa	88,570
9	Lufthansa	88,606	9	JAL	84,265	9	JAL	83,539
10	US Airways	75,728	10	US Airways	73,930	10	Qantas	75,134
11	Singapore	70,795	11	Southwest	71,591	11	Southwest	73,049
12	Southwest	67,924	12	Singapore	69,145	12	Singapore	71,118
13	Qantas	67,486	13	Qantas	67,894	13	Air Canada	69,417
14	All Nippon Group	62,592	14	Air Canada	67,016	14	US Airways	64,427
15	KLM	60,327	15	KLM	57,848	15	KLM	58,894
16	Cathay Pacific	47,153	16	All Nippon	56,904	16	All Nippon	52,973
17	Air Canada	57,374	17	Cathay Pacific	44,792	17	Cathay Pacific	49,041
18	TWA	43,791	18	Thai Int'l	44,039	18	Thai	48,513
19	Thai Int'l	42,395	19	Iberia	41,297	19	Koreqn	41,801
20	Alitalia	41,433	20	Korean	38,453	20	Iberia	40,470
21	Korean Air	40,532	21	Malaysia	38,313	21	Malaysia	36,897
22	Iberia	40,049	22	Alitalia	36,524	22	America West	31,983
23	Malaysia Airlines	37,939	23	Swissair	32,981	23	SAS Group	30,913
24	Swissair	34,246	24	TWA	31,848	24	Emirates	30,170
25	America West	30,753	25	America West	30,690	25	Alitalia	29,836

Rank	2003 Airline	2003 No. of RPKs (000,000)	Rank	2004 Airline	2004 No. of RPKs (000,000)	Rank	2005 Airline	2005 No. of RPKs (000,000)
1	American	193,604	1	American	209,708	1	American	222,634
2	United	167,970	2	United	185,407	2	Air France-KLM	185,709
3	Delta	143,896	3	Air France-KLM	168,998	3	United	183,862
4	Northwest	110,637	4	Delta	158,130	4	Delta	166,918
5	Air France	101,644	5	Northwest	118,108	5	Northwest	122,155
6	British Airways	100,850	6	British Airways	106,764	6	British Airways	111,859
7	JAL Group	93,847	7	Continental	105,786	7	Continental	110,003
8	Continental	92,662	8	Lufthansa Group	104,064	8	Lufthansa Group	108,185
9	Lufthansa Group	90,7208	9	JAL	102,354	9	US Airways Group	103,626
10	Southwest	77,225	10	Southwest	86,121	10	JAL Group	100,345
11	Qantas	77,225	11	Singapore	77,594	11	Southwest	97,097
12	Singapore	63,940	12	Qantas	73,630	12	Qantas	86,986
13	US Airways	60,792	13	Air Canada	69,887	13	Singapore	80,906
14	Air Canada	59,507	14	US Airways	65,170	14	Air Canada	75,243
15	KLM	57,368	15	Cathay Pacific	57,283	15	Cathay Pacific	65,110
16	All Nippon	50,182	16	All Nippon	55,735	16	China Southern	61,923
17	Thai	44,934	17	Thai	50,633	17	Emirates	59,299
18	Cathay Pacific	42,774	18	Emirates	48,749	18	All Nippon	58,481
19	Iberia	41,983	19	Air China	46,645	19	Korean	52,534
20	Emirates	40,000	20	Iberia	45,924	20	Air China	52,405
21	Korean	39,981	21	Korea	45,878	21	Thai	50,633
22	Malaysian	36,797	22	America West	37,544	22	Iberia	48,975
23	American West	34,247	23	China Southern	37,196	23	Malaysia	44,226
24	Air China	33,457	24	SAS Group	32,838	24	China Eastern	43,024
25	Virgin Atlantic	26,931	25	Malaysia	32,527	25	Alitalia	37,969

THE WORLD'S TOP 25 AIRLINES in FLEET SIZE

2000			2001			2002			2003			2004			2005		
	Airline	No of Aircraft	Rank	Airline	No of Aircraft	t	Airline	No of Aircraft	Rank	Airline	No of Aircraft	Rank	Airline	No of Aircraft	Rank	Airline	No of Aircraft
1	American	717	1	American	881	1	American	806	1	American	754	1	American	709	1	American	691
2	FedEx	662	2	FedEx	644	2	United	557	2	United	517	2	Delta	495	2	Delta	462
3	Delta	605	3	Delta	588	3	Delta	548	3	Delta	494	3	Northwest	433	3	Southwest	453
4	United	604	4	United	543	4	Northwest	446	4	Northwest	446	4	Southwest	424	4	United	404
5	Northwest	429	5	Northwest	444	5	Southwest	375	5	Southwest	375	5	United	418	5	SkyWest	395
6	US Airways	418	6	Southwest	355	6	Continental	358	6	Continental	358	6	Air France-KLM	355	6	NorthWest	365
7	Continental	372	7	Continental	345	7	FedEx	324	7	FedEx	324	7	Continental	337	7	Air France-KLM	363
8	Southwest	344	8	US Airways	342	8	US Airways	263	8	US Airways	263	8	FedEx	326	8	US Airways Group	356
9	British Airways	288	9	American Eagle	276	9	Air France	253	9	Air France	253	9	US Airways	278	9	Continental	350
10	American Eagle	261	10	British Airways	266	10	British Airways	237	10	ExpressJet	230	10	SAS Group	272	10	FedEx	338
11	Lufthansa	243	11	Air France	254	11	UPS	228	11	British Airways	228	11	American Eagle	256	11	ExpressJet	272
12	Air Canada	242	12	UPS	252	12	Lufthansa	226	12	UPS	224	12	ExpressJet	250	12	American Eagle	259
13	UPS	239	13	Air Canada	242	13	American Eagle	209	13	Lufthansa	218	13	British Airways	232	13	UPS Airlines	243
14	Air France	231	14	Lufthansa	236	14	Air Canada	204	14	JAL Group	215	14	UPS	232	14	British Airways	234
15	TWA	186	15	Continental Express	197	15	ExpressJet	188	15	American Eagle	212	15	Lufthansa	224	15	Lufthansa	233
16	Japan Airlines	171	16	Iberia	187	16	SkyWest	161	16	Air Canada	196	16	SkyWest	224	16	Japan Airlines	198
17	Continental Express	166	17	SAS	150	17	Alitalia	152	17	Skywest	192	17	JAL	209	17	China Southern	194
18	America West	164	18	Alitalia	144	18	Iberia	146	18	Alitalia	158	18	Air Canada	195	18	Air Canada	192
19	Iberia	159	19	America West	142	19	Atlantic Southeast	143	19	Comair	158	19	Comair	164	19	China Eastern	178
20	SAS	155	20	All Nippon	141	20	America West	142	20	Iberia	152	20	Iberia	158	20	Comair	169
21	Alitalia	146	21	Japan Airlines	141	21	Qantas	138	21	Atlantic Coast	150	21	Mesa	155	21	Iberia	159
22	All Nippon	140	22	Atlantic Southeast	132	22	SAS	138	22	American West	138	22	Alitalia	148	22	Mesa Airlines	157
23	Saudi Arabia	126	23	SkyWest	129	23	Atlantic Coast	136	23	Qantas	133	23	Atlantic Southeast	144	23	Alitalia	147
24	Comair	121	24	Saudi Arabia	128	24	All Nippon	135	24	All Nippon	130	24	America West	139	24	Air Canada Jazz	140
25	Aeroflot Russian	112	25	Korean	127	25	Swiss	134	25	Atlantic Southeast	130	25	ANA	136	25	ANA	139

THE WORLD'S TOP 25 AIRLINES in PASSENGERS

2000			2001			2002			2003			2004			2005		
Rank	Airline	Passengers (000)	Rank	Airline	Passengers (000)	Rank	Airline	Passengers (000)	Rank	Airline	Passengers (000)	Rank	Airline	Passengers (000)	Rank	Airline	Passengers (000)
1	Delta	105,723	1	Delta	104,943	1	American	94,144	1	American	88,241	1	American	92,565	1	American	98,098
2	American	86,280	2	American	78,178	2	Delta	89,868	2	Delta	84,245	2	Delta	86,901	2	Southwest	88,474
3	United	84,521	3	United	75,457	3	United	68,585	3	Southwest	74,787	3	Southwest	81,150	3	Delta	86,104
4	Southwest	63,678	4	Southwest	64,447	4	Southwest	63,046	4	United	66,100	4	United	70,879	4	Air France-KLM	69,159
5	US Airways	60,636	5	US Airways	56,114	5	Northwest	52,669	5	JAL	58,241	5	Air France-KLM	64,075	5	United	66,801
6	Northwest	58,722	6	Northwest	54,056	6	US Airways	47,168	6	Northwest	51,975	6	JAL	59,448	6	US Airways	63,998
7	All Nippon	49,887	7	All Nippon Group	49,306	7	Lufthansa Group	43,949	7	Lufthansa Group	45,400	7	Northwest	55,374	7	JAL	58,036
8	Continental	46,896	8	Continental	44,200	8	All Nippon	43,299	8	Air France	43,700	8	Lufthansa Group	50,901	8	Northwest	56,536
9	Lufthansa	41,300	9	British Airays	40,004	9	Continental	41,016	9	All Nippon	42,251	9	All Nippon	47,683	9	Lufthansa Group	51,255
10	Air France	39,204	10	Lufthansa	39,694	10	Air France	38,045	10	US Airways	41,263	10	Continental	42,743	10	All Nippon	49,609
11	British Airways	38,261	11	Air France	39,067	11	British Airways	34,445	11	Continental	38,903	6	US Airways	42,408	11	China Southern	44,118
12	Japan Airlines	33,857	12	JAL	37,183	12	JAL	33,631	12	British Airways	35,132	12	British Airways	35,680	12	Continental	42,822
13	Alitalia	26,697	13	Iberia	24,928	13	SAS Group	33,245	13	SAS Group	31,004	13	SAS Group	32,354	13	British Airways	35,634
14	TWA	26,392	14	Alitalia	24,926	14	Qantas	27,128	14	Qantas	28,884	14	China Southern	28,207	14	SAS Group	34,926
15	Iberia	24,543	15	SAS	23,244	15	Iberia	23,922	15	Iberia Group	25,097	15	Air Canada	27,600	15	Ryanair	33,384
16	Japan Air System	20,836	16	Air Canada	23,100	16	Air Canada	23,429	16	Korean	21,735	16	Iberia	26,692	16	Qantas	32,658
17	America West	19,954	17	TWA	22,214	17	Korean	22,172	17	Ryanair	21,372	17	Ryanair	26,583	17	Air Canada	30,000
18	Thai Int'l	18,038	18	Japan Air System	21,760	18	Alitalia	21,988	18	American West	20,047	18	Air China	24,500	18	EasyJet	29,600
19	Air Canada	17,655	19	Korean	21,638	19	China Southern	21,493	19	Air Canada	20,043	19	Qantas	24,270	19	China Eastern	29,377
20	China Southern	16,800	20	Qantas	20,193	20	Japan Air System	21,434	20	KLM	18,837	20	EasyJet	22,256	20	Air China	27,695
21	Malaysia Airlines	16,561	21	America West	19,576	21	America West	19,451	21	Easyjet	18,153	21	Alitalia	22,104	21	Iberia	27,436
22	KLM	16,234	22	China Southern	19,121	22	KLM	19,380	22	Air China	18,026	22	Korean	21,281	22	Alitalia	23,914
23	Singapore Airlines	14,874	23	Thai Int'l	18,271	23	Thai	18,729	23	Thai	17,301	23	American West	21,132	23	Korean	22,966
24	Swissair	14,238	24	Malaysia Airlines	16,745	24	Malaysia	16,208	24	China Southern	15,564	24	China Eastern	19,648	24	TAM	19,571
25	Alaska	13,525	25	KLM	15,995	25	Singapore	15,002	25	Malaysian	15,144	25	Thai	19,540	25	Thai	18,133

DIRECT AIRCRAFT OPERATING COSTS

EUROPEAN SHORT/MEDIUM RANGE

Aircraft Type	737-300	737-400	737-500	737-600	737-700	737-800
Study Range: 500nm						
Seats/Passengers	126/85.7	147/100	110/74.8	110/74.8	126/85.7	162/110.2
Block Time (HR)/Block Fuel (LB)	1.6/6,854	1.59/7,247	1.60/6,637	1.588/6,219	1.56/6,499	1.55/7,162
Operating Costs-per trip	$5,195	$5,469	$4,971	$4,831	$5,026	$5,511
US Dollars per NM	$10.39	$10.93	$9.94	$9.66	$10.05	$11.02
Relationship	BASE	5.30%	-4.30%	-7.00%	-3.30%	6.10%
US Cents per ASNM	8.24	7.44	9.03	8.78	7.97	6.8
Relationship	BASE	-9.80%	9.60%	6.50%	-3.30%	-17.50%

EUROPEAN SHORT/MEDIUM RANGE

Aircraft Type	757-200	757-300	767-200ER	767-300ER	767-400ER
Study Range: 1000nm					
Seats/Passengers	200/136	243/165.2	234/159.1	269/182.9	304/206.7
Block Time (HR)/Block Fuel (LB)	2.65/17,180	2.65/19,218	2.65/21449	2.64/22,640	2,63/25,007
Operating Costs-per trip	$11,029	$11,878	$13,551	$14,242	$15,132
US Dollars per NM	$11.02	$11.87	$13.55	$14.24	$15.13
Relationship	BASE	7.70%	22.90%	29.10%	37.20%
US Cents per ASNM	5.51	4.88	5.79	5.29	4.97
Relationship	BASE	-11.40%	5%	-4%	-9.70%

EUROPEAN LONG RANGE

Aircraft Type	747-100B	747-200B	747-300	747-400	777-200ER	777-300ER
Study Range: 3000nm						
Seats/Passengers	382/278.9	382/278.9	416/303.7	419/303.7	301/219.7	365/266.5
Block Time (HR)/Block Fuel (LB)	6.71/141,849	6.68/126,745	6.61/130,249	6.57/124,098	6.71/86,683	6.68/102,992
Operating Costs-per trip	$60,432	$57,111	$58,170	$53,556	$42,792	$47,951
US Dollars per NM	$20.14	$19.03	$19.39	$17.85	$14.26	$15.98
Relationship	BASE	-5.50%	-3.70%	-11.40%	-29.20%	-20.70%
US Cents per ASNM	5.26	4.98	4.66	4.26	4.73	4.39
Relationship	BASE	-5.50%	-11.60%	-19.20%	-10.10%	-17%

NOTES TO ABOVE:

Study Range	Figures are based on the range shown
Seats/Passengers	Total number of seats and passenger load factor used in calculation-typically 68-72%
Block Time	Fuel (LB)Total flight time and fuel used
Operating Costs-per trip	Operating costs per flight--crew/fuel/maintenance/landing fees/on-route charges
US Dollars per NM	US dollar cost per nautical mile
Relationship	% Increase/decrease from base aircraft in first column
US Cents per ASNM	US cents cost per total available seats on aircraft per nautical mile
Relationship	% Increase/decrease from base aircraft in first column
Engines	All aircraft are twins except for the 747 which has four engines

SOURCE: Boeing

WORLD SCHEDULED AIRLINES: SYSTEM FINANCIAL RESULTS, 1952-2005

(prior to 1998, excludes domestic operations within the CIS/USSR; for all years, U.S. data reflects majors and nationals only)

YEAR	OPERATING RESULTS				NET RESULTS	
	Revenues	Expenses	Profit/Loss	Margin	Profit/Loss	Margin
	(Millions, USD)			(%)	(Millions, USD)	(%)
1952	2,050	2,063	-13	-0.6	-45	-2.2
1953	2,314	2,317	-3	-0.1	-52	-2.2
1954	2,560	2,528	32	1.3	-33	-1.3
1955	3,025	2,947	78	2.6	-11	-0.4
1956	3,510	3,426	84	2.4	15	0.4
1957	3,971	4,012	-41	-1.0	-66	-1.7
1958	4,122	4,107	15	0.4	-46	-1.1
1959	4,805	4,700	105	2.2	-25	-0.5
1960	5,370	5,338	32	0.6	-97	-1.8
1961	5,795	5,913	-118	-2.0	-133	-2.3
1962	6,570	6,473	97	1.5	-27	-0.4
1963	7,153	6,824	329	4.6	106	1.5
1964	8,119	7,500	619	7.6	366	4.5
1965	9,347	8,460	887	9.5	534	5.7
1966	10,844	9,819	1,025	9.5	661	6.1
1967	12,488	11,575	913	7.3	609	4.9
1968	14,282	13,548	734	5.1	446	3.1
1969	16,431	15,557	874	5.3	409	2.5
1970	17,817	17,367	450	2.5	-10	-0.1
1971	20,116	19,507	609	3.0	138	0.7
1972	23,030	22,224	806	3.5	234	1.0
1973	27,438	26,243	1,195	4.4	434	1.6
1974	33,079	32,287	792	2.4	41	0.1
1975	38,309	37,579	730	1.9	-67	-0.2
1976	43,400	41,244	2,156	5.0	825	1.9
1977	50,344	47,715	2,629	5.2	1,656	3.3
1978	58,769	55,669	3,100	5.3	2,412	4.1
1979	70,755	70,019	736	1.0	588	0.8
1980	87,676	88,311	-635	-0.7	-919	-1.0
1981	92,992	93,684	-692	-0.7	-1,150	-1.2
1982	93,240	93,400	-160	-0.2	-1,300	-1.4
1983	98,300	96,200	2,100	2.1	-700	-0.7
1984	105,400	100,300	5,100	4.8	2,000	1.9
1985	112,200	108,100	4,100	3.7	2,100	1.9
1986	124,600	120,000	4,600	3.7	1,500	1.2
1987	147,000	139,800	7,200	4.9	2,500	1.7
1988	166,200	156,000	10,200	6.1	5,000	3.0
1989	177,800	170,200	7,600	4.3	3,500	2.0
1990	199,500	201,000	-1,500	-0.8	4,500	-2.3
1991	205,500	206,000	-500	-0.2	-3,500	-1.7
1992	217,800	219,600	-1,800	-0.8	-7,900	-3.6
1993	226,000	223,700	2,300	1.0	-4,400	-1.9
1994	244,700	237,000	7,700	3.1	-200	-0.1
1995	267,000	253,500	13,500	5.1	4,500	1.7
1996	282,500	270,200	12,300	4.4	5,300	1.9
1997	291,000	274,700	16,300	5.6	8,550	2.9
1998	295,500	279,600	15,900	5.4	8,200	2.8
1999	305,500	293,200	12,300	4.0	8,500	2.8
2000	328,500	317,800	10,700	3.3	3,700	1.1
2001	307,500	319,300	-11,800	-3.8	-13,000	-4.2
2002	316,000	310,900	-4,900	-1.6	-11,300	-3.7
2003	321,800	323,300	-1,500	-0.5	-7,560	-2.3
2004	378,800	375,500	3,300	0.9	-5,570	-1.5
2005	413,300	409,000	4,300	1	-3,200	-0.8

SOURCE: ATA/ICAO

BIBLIOGRAPHY

Adams, H.W. The Inside Story: The Rise and Fall of Douglas Aircraft Toledo Publishing, Batangas, Philippines, 2000.

Aris, S. Close to the Sun: How Airbus challenged America's domination of the skies Aurum Press, London, UK, 2002.

Barlay, S. Cleared for Take-off: Behind the Scenes of Air Travel Kyle Cathie Ltd., London, UK, 1994.

Bauer, E.E. Boeing: The First Century TABA Publishing Inc., Washington, USA, 2002.

Brooks, P. W. "Transport Aircraft Development: Parts 1, 2 and 3" AIR Pictorial, Profile Books Ltd., London, UK, May, June and July 1985 issues.

Davies, R.E.G. Airlines of the United States since 1914 Putnam and Co. Ltd., London, UK, 1972.

Davies, R.E.G. A History of the World's Airlines Oxford University Press, London, 1967.

Evans, J. How Airliners Fly: A Passenger's Guide Airlife Publishing Ltd., England, 2002.

Francillon, R.J. McDonnell Douglas Aircraft since 1920 Putman and Co. Ltd., London, UK, 1979.

Gann, H. Douglas DC-6 and DC-7 Specialty Press, MN, USA, 1999.

Greenwood, J.T. (ed.) Milestones of Aviation Hugh Lauter Levin Assoc. Inc., New York, USA, 1989.

Gunn, J. Challenging Horizons: Qantas 1939-1954 University of Queensland Press, St Lucia, Australia, 1990.

Gunn, J. High Corridors: Qantas 1954-1970 University of Queensland Press, St Lucia, Australia, 1990.

Gunn, J. The Defeat of Distance: Qantas 1919-1939 University of Queensland Press, St Lucia, Australia, 1988.

Gunston, B. "A Tale of Two Rivers" AIR International, Key Publishing Ltd., Stamford, UK, April 2001 issue.

Hardy, M.J. The Lockheed Constellation David and Charles (Holdings) Ltd., Devon, UK, 1973.

Holden, Henry M. The Douglas DC-3 Airlife Publishing Ltd., England, 1991.

Ingells, D.J. L-1011 TriStar and the Lockheed Story Aero Publishers Inc., Cal., USA, 1973.

Ingells, D.J. 747: Story of the Boeing Super Jet Aero Publishers Inc., California, USA, 1970.

Jarrett, P. Modern Air Transport: Worldwide Air Transport from 1945 to the Present Putnam Aeronautical Books, London, UK, 2000.

Morrison, W.H. Donald W. Douglas: A Heart with Wings Iowa University Press, USA, 1991.

Newhouse, J. The Sporty Game: The high-risk competitive business of making and selling commercial airliners Alfred A. Knopf, New York, USA, 1982.

Norris, G. and Wagner, M. Boeing 777: The Technological Marvel MBI Publishing Co., WI, USA, 2001.

Orlebar, C. The Concorde Story: Fifth Edition Osprey Publishing, Oxford, UK, 2002.

Pearcy, A. Douglas Propliners DC-1 - DC-7 Airlife Publishing Ltd., Shrewsbury, UK, 1995.

Proctor, J. Convair 880 & 990 World Transport Press, Florida, USA, 1996.

Pugh, P. The Magic of a Name: The Rolls-Royce Story. Part Three: A Family of Engines Icon Books Ltd., Cambridge, UK, 2002.

Ramsden, J.M. The Safe Airline MacDonald and Jane's Publishers Ltd., London, UK, 1976.

St. John Turner, P. Pictorial History of Pan American World Airways Ian Allan Ltd., London, UK, 1973.

Stackhouse, J. ...from the dawn of aviation: The Qantas Story 1920-1995 Focus Publishing, Double Bay, Australia, 1995.

Stringfellow, C.K. and Bowers, P.M. Lockheed Constellation: Design, Development, and Service History of all Civil and Military Constellations, Super Constellations, and Starliners Motorbooks International, Wisconsin, USA, 1992.

Szurovy, G. Classic American Airlines MBI Publishing Co., Wisconsin, USA, 2000.

Upton, J. Lockheed L-1011 Tristar Specialty Press, MN, USA, 2001.

Veronico, N.A. Boeing 377 Stratocruiser Specialty Press, MN, USA, 2001.

Waddington, T. Douglas DC-8: Volume 2 World Transport Press, Florida, USA, 1996.

Waddington, T. McDonnell Douglas DC-9 World Transport Press, Florida, USA, 1998.

Whitford, R. "Fundamentals of Airliner Design: Parts 1, 2, 3, 9, 10, 12" AIR International Key Publishing Ltd., Stamford, UK. February 2001, April 2001, June 2001, June 2002, August 2002, January 2003, issues.

Yenne, B. Classic American Airliners MBI Publishing Co., MN, USA, 2000.

Yenne, B. The Story of the Boeing Company AGS BookWorks, CA, USA, 2003.

Yule, P. The Forgotten Giant of Australian Aviation: Australian National Airways Hyland House Publishing Pty Ltd., Flemington, Australia, 2001.